MW00332881

The Subjectivity of Scientists and the Bayesian Approach

S. James Press
Emeritus Distinguished Professor of Statistics
University of California, Riverside

Judith M. Tanur
Distinguished Teaching Professor, Emerita
State University of New York, Stony Brook

Dover Publications, Inc.
Mineola, New York

Copyright

Copyright © 2001 by S. James Press and Judith M. Tanur
All rights reserved.

Bibliographical Note

This Dover edition, first published in 2016, is an unabridged republication of
the work originally published in 2001 by John Wiley & Sons, Inc., New York.

Library of Congress Cataloging-in-Publication Data

Names: Press, S. James. | Tanur, Judith M.
Title: The subjectivity of scientists and the Bayesian approach / S. James Press
and Judith M. Tanur.
Description: Dover edition. | Mineola, New York : Dover Publications, Inc.,
2016. | Originally published: New York : John Wiley & Sons, Inc., 2001.
Includes bibliographical references and index.
Identifiers: LCCN 2015038004 | ISBN 9780486802848 | ISBN 0486802841
Subjects: LCSH: Research—Methodology. | Bayesian statistical decision
theory. | Subjectivity.
Classification: LCC Q180.55.M4 .P73 2016 | DDC 507.2—dc23 LC record
available at http://lccn.loc.gov/2015038004

Manufactured in the United States by RR Donnelley
80284101   2016
www.doverpublications.com

To our spouses and children

The Reverend Thomas Bayes

Contents

Preface

We expect that both professional scientists and the general public will benefit from reading this book, not only to become better informed, but also for pure enjoyment. Our point of view in this book is that knowledge is accumulated by continually updating our understanding of a phenomenon through the merging of current findings with previous information. We believe that this is how scientists have worked informally in the past, and how scientists might work more formally in the future.

Nonscientists will benefit from reading the book by gaining an increased understanding of how scientists have really worked—an understanding that is different from what they've traditionally been taught. Nonscientists reading this book will surely improve their understanding of the many ways that personal beliefs and biases find their way into supposedly objective scientific results and conclusions. After having read the book, laypeople will be likely to review reported results of scientific studies with a more critical, questioning, and skeptical eye, an important stance in today's increasingly complex society. [See Nelkin (1995) for additional discussion of how the personal biases of scientists affect the way they sometimes communicate their results to the public.]

This book will also be interesting and useful reading for science specialists. They will be reminded of how important it is to treat the data from their experiments with great care and respect, and they will see how their educated scientific beliefs can be introduced into their research in ways that they can use profitably without compromising their results. We believe that most specialists are not likely to have seen the common threads of tenacity of belief in the face of contradictory data that have run through the work of so many of the most famous scientists in history, even such giants as Sir Isaac Newton, Galileo Galilei, and Albert Einstein. These icons of science have had such keen awareness and understanding of the basic principles underlying their disciplines that their intuition and scientific judgment generally helped to lead them to correct conclusions.

Statisticians will find that the book integrates the world of inference into the general framework of science in a way that demonstrates the interdisciplinary nature of their field. They will also be reminded of the broadly pervasive role played by subjectivity.

All readers are likely to benefit from the final chapter. It is somewhat more technical than the earlier chapters because it addresses the subject of *Bayesian statistical analysis* of scientific data. The subject is treated at an introductory, elementary level that is designed to provide an appreciation of how specialists and nonspecialists alike might benefit from adopting such a viewpoint. We believe that this chapter, especially in the context of the rest of the book, makes the volume a useful supplementary text for an undergraduate or graduate course in Bayesian analysis.

This book has had a gestation period of over 12 years. During that time we have learned to work together and help each other's understanding of the issues we were tackling. But we have been helped by others as well. Discussions with colleagues, friends, and family members and the suggestions they made helped us to give more careful thought to the fundamental issues discussed in this book. We are grateful to Gordon Kaufman, John Kimmel, Frank Lad, Teddy Seidenfeld, Arnold Zellner, and several anonymous referees for helpful suggestions for improving the manuscript. We are especially grateful to Rachel Tanur for producing the drawings of the scientists. We are also grateful for the financial support and encouragement given to us by the Department of Statistics, University of California, Riverside, and the Department of Sociology, State University of New York, Stony Brook.

Oceanside, California   S. JAMES PRESS
Montauk, New York   JUDITH M. TANUR
August 7, 2000

REFERENCE

Nelkin, Dorothy (1995). *Selling Science: How the Press Covers Science and Technology*, rev. ed., New York: W.H. Freeman.

CHAPTER 1

Introduction

This is a book about science, about scientists, and about the methods that scientists use. We will show that the most famous scientists in history have all used their hunches, beliefs, intuition, and deep understanding of the processes they studied, to one extent or another, to arrive at their conclusions. The reader will see that the oft-expressed notion that science is "objective" is only partially true; in fact, science is really a combination of both subjective and objective views.

In this book we tell the stories of 12 of the most famous scientists throughout history, from Aristotle, the philosopher who lived during the era of the ancient Greeks, to Albert Einstein, who lived during the twentieth century. In each case, we discuss how the scientist's preconceived beliefs informally influenced his or her scientific conclusions.

In Chapter 2 we explain that we did not choose these particular 12 scientists to study; they were selected for us. In Chapter 3 we tell the stories of five other very celebrated cases of famous scientists who may have stepped over the lines of acceptable scientific practice. Such overstepping arose either because their convictions about the correctness of their ideas led them to see or accept as accurate only what their theory predicted, or from their zeal to convince the world of the scientific merits of their work.

A major portion of the book is devoted to the stories of our 12 most famous scientists in history (Chapter 4). We examine the lives of these people, their scientific contributions, and the ways in which they used their beliefs together with the results of their scientific experiments to carry out their research. In a final section of Chapter 4 we conjecture about what these scientists might have done had modern methods been available to them for combining their preconceptions about the processes they were studying with their experimental data.

Finally, in Chapter 5, we examine how subjectivity is being used in science in modern times. That discussion focuses on *Bayesian statistical science*. But first, in this introductory chapter, we provide some definitions and background for what is to come.

1.1 SUBJECTIVITY AND OBJECTIVITY

Because we will be using the words *subjectivity* and *objectivity* frequently in this book, we must first explain how we are using these terms. We intend to use them in several ways. But we will always be thinking about observational data, about the distinction between beliefs held by a scientist about a phenomenon prior to collecting the data, and beliefs held by the scientist after the data have been collected and analyzed.

In common usage, certain entities are seen to have only a *subjective reality*, in that those entities are constructs based on views and beliefs that were formed out the human mind. On the other hand, the term *objective reality* is used for entities that exist outside the minds of individuals, in that they exist in the world regardless of whether a person perceives them. For example, the opinions people hold about a political or social issue are personal beliefs that have only a subjective reality, whereas our starting point with respect to external reality is that the moon, the sun, and the planets would all exist regardless of whether or not human beings perceived them. But objective entities need not be corporeal. For example, all the scientific laws that govern the behavior of the physical world would exist regardless of their codification by human beings or humans' belief in them.

The crux of our usage makes these terms specific to the scientific endeavor. When anything is observed or measured by human beings, human perception and sensory mechanisms are involved, and the resulting observations that are collected then depend on subjectively based distortions of objective reality. Further, after entities in the objective world are observed (perceived) by a human being, either through the senses and/or assisted by measuring instruments, the resulting measurements (called *data*) are interpreted by a human being in a subjective way that reflects the person's own experience, understanding, and preconceived notions and beliefs about the entities, the object, phenomenon, or construct being measured. The interpretation of such data is also affected by the person's state of mind and state of senses at the time of the interpretation. We use the term *subjective*, or *subjectivity*, to refer to pre-existing views or beliefs about entities that influence both the gathering of data and their interpretation.

Broadening our interpretation of the term, we will also use *subjectivity* to mean a person's intuition, belief, and understanding about some proposition or hypothesis prior to that person collecting observational data that bear on the proposition, or prior to that person obtaining information about the hypothesis. Because the views about the hypothesis that we are talking about are those that a particular person held prior to that person having collected data, or they are his or her views prior to being told about the values of data that have been collected, and because these views may well differ across individuals, we call those views or beliefs *subjective*.

To define *objectivity* in an experimental context, we appeal to a (grossly oversimplified) description of the old-fashioned textbook image of how science proceeds and how scientists behave. In this image, science and scientists are *objective* in the following sense. A hypothesis is developed (the passive voice here is used to signify that little attention is paid to the origin of the hypothesis) and the scientist designs a study to test this hypothesis. After data gathering, whether by designed experiment or by carefully carried out observation in a nonexperimental setting, the scientist dispassionately evaluates the results and their implications for the hypothesis. If the results support the hypothesis, the scientist writes up the study for publication; if they do not, the scientist, again dispassionately, abandons the hypothesis as being wrong, and either revises the hypothesis in light of the new findings and repeats the cycle, or goes on to other concerns.

The subjectivity that we intend to demonstrate that scientists use routinely is both less and more than the opposite of this kind of mythical objectivity. We surely do not suggest that scientists ignore the objective reality that is "out there," nor that they ignore the guidance offered by the results of their investigations. And surely historians of science have long since made it clear that scientists are considerably more human than our oversimplified portrait of dispassionate automaton would indicate. Scientists care about their work, care about their results, and even care about the recognition that will come to them with successful discoveries. These factors often drive the way in which they carry out their research and the way they report it to the world. Despite these factors, some observers of scientific methodology still describe it as "objective."

Strongly held personal beliefs and hypotheses about scientific phenomena, the stuff of what we refer to as subjectivity, are sometimes so strongly held that a scientist will, under their influence, announce confirmatory results from experiments not yet carried out. Such a practice is clearly fraudulent, as is introducing major alterations of data from actual experiments to make them conform to a subjectively held theory. But we need to understand that the dividing line between normal practice and fraud is sometimes a fine one. For example, normal statistical analysis of scientific data often requires the analyst to decide to drop some data points because they lie so far away from what is expected that they seem to be aberrations or mistakes rather than meaningful data that should be considered along with the bulk of the other observations (such data points are sometimes called *outliers*), or because they are poorly measured.

But what we mean by the subjectivity of scientists is deeper than these understandably human traits. Some scientists seem to be particularly opinionated and stubborn, not unlike some nonscientists. These scientists develop hypotheses based on strongly held, preconceived notions of how the world operates, and they sometimes (usually, unconsciously) design studies to prove these notions rather than merely to test them. In cases in which the results of their investigation are contrary to what the hypothesis would predict, a scientist is sometimes more likely to doubt that the data are accurate than to conclude that the theory is incorrect. Such a scientist will redesign the study and persist in trying to find data that prove the hypothesis, sometimes for years, sometimes for a lifetime.

1.2   SUBJECTIVE AND OBJECTIVE PROBABILITY

There is another important and related sense in which we use the terms *subjective* and *objective* in this book; it relates to their meanings in relation to *subjective* and *objective probability*. This use of the terms is discussed in great detail in Chapter 5. For the time being, we merely state that *subjective probability* refers to an individual scientist's degree of belief about the chance of some event occurring. *Objective probability* refers to the mathematical or numerical probability or chance of some event occurring.

1.3   RATE-OF-DEFECTIVES EXAMPLE

The definition of objectivity employed by some philosophers requires that the statement be testable by anybody. Subsumed in this position is the notion that everyone must have the same interpretation of observational data. But we believe that such uniformity of interpretation is rarely the case. We believe that different people come to observational data with differing preexisting views that induce them to construe the interpretation of the data somewhat differently, depending on their preconceived biases. It will be illuminating to begin our discussion of subjectivity in scientific methodology with an example that illustrates how different scientific observers, usually unwittingly, bring their own beliefs and biases—their subjectivity—to bear on the interpretation of scientific data. We see in the example below how different observers of the same data can proceed very differently and thus come away with very different interpretations of them. We call this illustration the *rate-of-defectives example*, and we refer to it later.

Before we present the example, however, we quote Ian Hacking (1965, p. 217), who noted that "[o]ne of the most intriguing aspects of the subjectivist theory, and of Jeffreys' theory of probability, is use of a fact about Bayes' theorem to explain the possibility of learning about a set-up from a sequence of trials on it. The fact seems to explain the possibility of different persons coming to agree on the properties of the set-up, even though each individual starts with different prior betting rates and different initial data." Here Hacking refers to the comforting fact that although different observers of the same data may have differing interpretations of them, eventually, with a sufficiently large number of trials, the differing prior views of different observers about the same data will generally disappear as the data begin to dominate all personal views about the underlying process.

Let us suppose that you collect 100 observations from an experiment. We can refer to these observations as *data points*. You then send these data to five scientists located in five different parts of the world. All five scientists receive the same data set, that is, the same 100 data points. (Note that for purposes of this example, the subjectivity involved in originally deciding what data to collect and in making the original observations themselves is eliminated by sending the same "objective" data to all five scientists.) Should we expect all five of the scientists to draw the same conclusions from these data?

Our answer to this question is a very definite "no." But how can it be that different observers will probably behave very differently when confronted with precisely the same data? Part of the thesis of this book is that the methodology of science requires that inferences from data be a mixture of both subjective judgment (theorizing) and objective observation (empirical verification). Thus, even though the scientists all receive the same observational data, they come to those data with differing beliefs about what to expect and about how to proceed. Consequently, some scientists will tend to weight certain data points more heavily than others and consider differing aspects of the data as more consequential. Different scientists are also likely to weight experimental errors of measurement differently from one another. Moreover, scientists may decide to carry out formal checks and statistical tests about whether the phenomenon of interest in the experiment was actually demonstrated (to ask how strongly the claimed experimental result was supported by the data). Such tests are likely to have different results for different scientists, because different scientists will bring different assumptions to the choice of statistical test. More broadly, scientists often differ on the mathematical and statistical models they choose to analyze a particular data set, and different models usually generate different conclusions. Different assumptions about these models will very often yield different implications for the same data.

These ideas that scientists can differ about the facts are perhaps surprising to some of us. Let us return to our 100 observations and five scientists to give a very simple and elementary example, with the assurance that analogous arguments will hold generally for more realistic and more complicated situations.

Hypothetically, suppose there is a special type of machines that produces a certain component that we can call a "groove joint," a component required for the hard drives of desktop computers. It is common knowledge in the computer industry that because groove joints are very difficult to fabricate, such machines generally produce these components with about a 50% defective rate. That is, about half the groove joints produced by a given machine will have to be discarded because they are defective. The machines that produce groove joints are very expensive.

The hypothetical South Bay Electronics Company suspects that its newly purchased machine may be producing groove joints at a defective rate different from the industry norm, so it decides to test the rate at which the new machine produces defectives. It first fabricates 100 groove joints on the new machine (each groove joint is fabricated independently of every other groove joint) and examines each one to classify it as "good" or "defective." South Bay records the sequence of 100 groove joints produced by the new machine as: G, G, D, . . . , with "G" representing "good" and "D" representing "defective." The company finds that there were 90 defective groove joints in this batch, but the quality control staff still doesn't know the long-run rate of defectives for their machine. They decide to send the results representing the 100 tested groove joints to five different scientists to ask them for their own estimates of the long-run rate of defectives for this machine. We shall call this long-run rate of defectives p, bearing in mind that p can be any number between zero and one.

The sequence of G's and D's are the data you send to the five scientists (three women and two men) in five different locations around the world to see how they interpret the results. You tell them that you plan to publish their estimates of the long-run rate of defectives and their reasoning behind their estimates in a professional scientific journal. Thus their reputations are likely to be enhanced or to suffer in proportion to their degrees of error in estimating p. As we shall see, it will turn out that they will all have different views about the long-run rate of defectives after having been given the results of the experiment.

Scientist 1 is largely a *theorist* by reputation. She thinks that $p = 0.5$ no matter what. Her line of reasoning is that it just happened that 90% of the first 100 groove joints were defective, that there was a "run" of "defectives." Such an outcome doesn't mean that if the experiment were to be repeated for another 100 trials (produce yet another 100 groove joints) the next 100 trials wouldn't produce, say, 95 defectives, or any other proportion of defectives. Scientist 1 has a very strong preconceived belief based upon theory that groove joints are about equally likely to be good or defective ($p = 0.5$) in the face of real data that militate against that belief. For her, unless told otherwise, all machines produce defective groove joints for roughly half of their output, even if many runs of many defectives or many good groove joints just happen to occur.

Scientist 2 has the reputation for being an *experimentalist*. He thinks that $p = 0.9$, because that is the proportion of defectives found in the batch. (This estimate of p is what statisticians call the *sample estimate* (and under certain conditions, they also call it the *maximum likelihood estimate*). Scientist 2's definition of the best estimate available from the data is the fraction of defectives actually obtained. While scientist 1 believed strongly in theory, scientist 2 is ready to abandon theory about the production of defective groove joints in favor of strong belief in data, regardless of theory.

When he actually carries out experiments in his research, which is only occasionally, scientist 3 is an extremely *thorough and careful experimentalist*. He carefully examines the actual sequence of outcomes and decides to ignore a run of 50 straight defectives that occurred, feeling that such a run must be a mistake in reporting since it is so unlikely. This reduces the number of available data points (the sample size) from 100 down to 50, and out of those 50, 40 were defective. So his conclusion is that $p = 40/50 = 0.8$. Scientist 3 has taken the practical posture that many scientists take of weighting the observations so that some observations that are believed to be errors are discarded or down-weighted in favor of those thought to be better measurements or more valid in some experimental sense.

Scientist 4 is a *decision theorist*, driven by a need to estimate unknown quantities on the basis of using them to make good decisions. Such a scientist would be interested in minimizing the costs of making mistakes, and might perhaps decide that overestimating p is as bad as underestimating it, so the costs should be the same for making these two types of errors. Moreover, he wants to select his estimator of p to enhance his reputation, and you have told him that a correct estimate will do just that. So he decides to select his estimator such that the cost of being wrong will be as small as possible, regardless of the true value of p. (In the theory of decision making, under such circumstances he would often adopt what is called a *quadratic loss function*.) If scientist 4 were to adopt the subjective belief that all values of p are equally likely and then used a result from probability theory (called *Bayes' theorem*; see discussion of this theorem in Chapter 5), his resulting estimate of p would be $p = 91/102 = 0.892$.

Scientist 5 may be said to be *other-directed*. She badly wants the recognition from her peers that you said would depend on the accuracy of her estimate. She learns from you the names of the other four scientists. She then writes to them to find out the estimates they came up with (and for their own reasons, they send her their results). Having obtained the results of the other scientists, scientist 5 decides that the best thing to do would be to use their average value. (Robert Millikan, winner of a Nobel Prize in Physics, did similar averaging of earlier reported results—from entirely separate experiments in that case—in his oil-drop experiment to measure the charge on an electron; see Section 3.4.) So for scientist 5, the estimate of $p = (0.5 + 0.9 + 0.8 + 0.892)/4 = 0.773$. Or perhaps, learning that scientist 1 made an estimate of p that was not data dependent, scientist 5 might eliminate scientist 1's estimate from her average, making the subjective judgment that it was not a valid estimate. Then scientist 5 would have an estimate of $p = (0.9 + 0.8 + 0.892)/3 = 0.864$. Scientist 5's strategy is used by many scientists all the time, so that the values they propose will not be too discrepant with the orthodox views of their peers.

So the five scientists came up with five different estimates of p starting with the same observed data. All the estimates are valid. Each scientist came to grips with the data with a different perspective, using somewhat different procedures, and with a different belief about p.

The values of p found by the five scientists thus are as follows:

Scientist	1	2	3	4	5
Estimated Value of p	0.500	0.900	0.800	0.892	0.773 or 0.864

But what is the "true" value of p?

In empirical science we never know with certainty what is true and what is not. We have only educated beliefs and the results of observations and experiments. But note that with the exception of scientist 1, who refuses to be influenced by the data, all the scientists agree that p must be somewhere between 0.773 and 0.900, agreeing that the machine is turning out defective groove joints at a long-run rate of greater than 0.5.

Suppose that the machine had produced 1000 groove joints instead of the 100 groove joints we just considered, but with analogous results of 90% defectives. Would this have made any difference to the scientists? Well perhaps it might have, for example, to scientist 1, because people differ as to the point at which they will decide to switch positions from assuming that the machine produces defectives at typical rates despite the experimental outcome results, to a position in which they are willing to assume that the machine has a higher long-run defective rate. One scientist may switch after 9 defectives out of a batch of 10 groove joints, another after 90 out of 100, whereas another may not switch until there are perhaps 900 defectives out of a batch of 1000 groove joints, or might insist on an even more extensive experiment with even stronger evidence. In any case the scientists may still differ in their views about the long-run rate of defectives for the machine for a very wide variety of reasons, only a few of which are mentioned above.

1.4   DIVERSITY IN THE SCIENTIFIC METHODOLOGY PROCESS

The rate-of-defectives example we just discussed was very elementary, but it was science nevertheless. Similar interpretive and methodological issues arise in all branches of the sciences. To examine the issues more broadly, we step back to the origins of science itself, and to the structure, development, and applications of various scientific methodologies.

Wolpert (1992, p. 35) suggests that "unlike technology or religion, science originated only once in history, in Greece." He asserts that "Thales of Miletos who lived in about 600 B.C., was the first we know of who tried to explain the world not in terms of myths but in more concrete terms, terms that might be subject to verification." (Miletos is a city on the Aegean coast of what is now Turkey; at that time the city was considered part of Greece.) This was the beginning of what has come to be known as *science*. (In fact, it is claimed that Thales predicted the total solar eclipse of 585 B.C., and was responsible for proving five theorems of plane geometry, including the fact that an angle inscribed in a semicircle must be a right angle.)

There is no unique definition for the term *scientific method*, a term discussed at length by Francis Bacon, Immanuel Kant, and others. But we can try to characterize the scientific methodology process. We will see that scientists have varied greatly in their methodological approaches to research, so that there has not really been a single, acceptable approach regarded as the "correct" path to scientific achievement. In the following, we use the scientists studied in Chapters 3 and 4 to illustrate these ideas.

Alexander von Humboldt (Section 4.7), a founding father of the field of physical geography, mostly observed and carefully recorded his observations. In some instances, he also hypothesized possible explanations about the mechanisms underlying his observations and showed how the observations fit these hypotheses. He was the quintessential observer. But as physical geography was not viewed in earlier years as a field that required experimentation, he rarely conducted experiments. So the vast bulk of his beliefs about the possible causes of the phenomena he observed were totally subjective, and not based on experimental verification.

Albert Einstein (Section 4.13), who propounded the theory of relativity, mostly hypothesized about his observations, relying heavily on his profound understanding of physics and on the implications of mathematical models he developed for describing physical phenomena to make verifiable predictions about the universe. But he personally almost never carried out experiments that might have demonstrated the validity of his mathematical inferences and subjective speculations about the world. Other scientists, sometimes many decades later, did ultimately check and verify many of Einstein's predictions empirically.

Charles Darwin (Section 4.9), the biologist who propounded the theory of evolution, spent years in the process of observing and recording data meticulously, but then he devoted the next 23 years of his life to theorizing and hypothesizing, trying to generalize from his observations. Predictive mechanisms for the process of biological inheritance were developed by others, such as Gregor Mendel (discussed in Section 3.3).

Sigmund Freud (Section 4.11) treated many people with emotional and mental problems and recorded his subjective observations case by case. He then spent many years trying to generalize his subjective beliefs about his patients to synthesize a theoretical framework from his case studies.

Sir Isaac Newton (Section 4.5), perhaps the greatest of all scientists of all time, observed, hypothesized, developed the mathematics he needed to create mathematical models for use in quantitative prediction about the phenomena he wanted to study, built scientific equipment to improve the quality of his observations, carried out experiments designed to test his hypotheses, and then made and checked his predictions about the world. But as will be seen in Section 4.5, his strong beliefs about his theories may have led him to misrepresent some of his experimental results relating to the speed of sound.

Marie Curie (Section 4.12) was another quintessential experimenter. She carried out experiments, laboriously and painstakingly, extending over many years, experiments designed to isolate the new element, radium. Her subjective beliefs that radium was indeed an element and could be isolated from pitchblende gave her the patience to continue searching until she gathered a sufficient quantity of the precious element to establish its atomic properties.

Some scientists did not attempt to establish quantitative relationships. Rather, they carried out demonstrations to show that certain types of effects existed. For example, William Harvey (Section 4.4) provided a demonstration of the circulation of blood.

So sometimes a scientist merely makes observations; in other cases, the scientist observes and then hypothesizes but does not follow up with experiments; in other cases, the scientist observes, hypothesizes, experiments, or demonstrates, and then perhaps rehypothesizes, and reexperiments, often cycling through the investigative process many times. There is a complete spectrum of methodological behaviors followed by scientists. Most scientists labor in the vineyards to extend other people's theories and end up falling somewhere in between the extremes of the methodological investigatory spectrum. There is no set rule about what procedure every scientist follows. Francis Bacon (1561–1626), a philosopher, attempted to establish some rules about *proper* scientific methodology. His approach was experimental, relying on inductive logic. He was not convinced that mathematics should play an important role in the scientific process, so his method was qualitative. Bacon's approach to scientific methodology was not really adopted until the nineteenth century and was quite limited relative to the broader spectrum of procedures actually adopted by scientists over the centuries.

1.5 CREATIVITY IN SCIENCE

It is most usual that scientific advance starts with an idea or a question in a scientist's head about some phenomenon. [Occasionally, a scientific advance comes about serendipitously, in that the scientist notices something unanticipated but realizes that it is important and thus discovers some unexpected effect; see, for example, Roberts (1989). We confine our attention to more usual, nonserendipitous discoveries.] But how does such an idea or question arise? A scientist observes the surrounding world, as do we all, but when the scientist has observed some phenomenon, perhaps many times, a trained curiosity prompts the asking of such questions as: I wonder what caused that to happen? I wonder what would happen differently if the conditions were changed somewhat? Is the response directly proportional to the stimulus? I wonder what the role of randomness was in that phenomenon?

This process, stimulated by scientific curiosity, of raising questions about the mechanism that generates some real-world effect, sets the scientist to theorizing and formulating hypotheses about the phenomenon. The process demands that the scientist be creatively subjective, and think nonlinearly, much as the artist, composer, or creative writer behaves. The scientist must be able to leap to tentative conclusions about what to expect under certain circumstances, without having yet observed the phenomenon under those circumstances and without passing through the appropriate logical steps (linear thinking) required to arrive at deductive conclusions. This subjectively creative process is fundamental to how science is carried out. The scientist begins searching through a lifetime of personal experience, knowledge, and understanding of related phenomena. A review of the published literature on related topics reveals what is already known about this subject. If the full answers are not known, the exploratory path is open and the scientist begins to pursue the topic with vigor. Thus, the scientist proceeds with the idea, with speculation about it, and the scientific research process begins. Within this creative process, the scientist is most often driven by his or her own biases, intending to find preconceived results (results within a certain range) in the experiment. However, as Kuhn (1962, p. 35) pointed out: "[T]he range of anticipated, and thus of assimilable, results is always small compared with the range that imagination can conceive. And the project whose outcome does not fall in the narrower range is usually just a research failure, one which reflects not on nature but on the scientist."

1.6 THOUGHT (IMAGINARY) EXPERIMENTS VERSUS PHYSICAL EXPERIMENTS

Scientific research usually involves both theory and experiment. That is, a scientist, after theorizing about the mechanism that is involved in a phenomenon, typically carries out experiments that will bear on the mechanisms postulated to be involved. Good experiments are designed to demonstrate whether the hypotheses about observed phenomena were correct. Usually, not all of a scientist's ideas are verified by an experiment or series of experiments, and modifications of the original beliefs must be made and new experiments mounted to test the modified beliefs.

The scientist who merely hypothesizes and does not experiment is being strictly speculative; in such a case, subjective beliefs and the logic (perhaps mathematical projections or predictions) that ties them together are all that are available to justify scientific conclusions. These hypothesized beliefs may be arrived at by "thought experiments" in which the scientist reasons about the phenomenon, and based on a special understanding of how it works, tries to conclude what is likely to happen under various conditions or scenarios. Without empirical verification of the logically derived hypothetical conclusions, the scientist is likely to be wrong, at least in part. In some unusual and remarkable cases, the scientist hypothesizes correctly about the likely outcome of experiments that *could* be carried out but *haven't been yet*. That is, such a scientist correctly predicts what will happen under various conditions. One of the characteristics of the *outstanding* scientist is that these subjective judgments about the phenomenon under study turn out to be correct, even though the conclusions reached do not appear to have been arrived at through logical, step-by-step reasoning, experimental verification, or any combination thereof. There are many instances of this interesting, characteristic trait of the 12 outstanding scientists on whom we focus in Chapter 4. There, we summarize the lives, the scientific achievements, and the methodological approaches to science adopted by these very famous scientists.

Some scientists make the decision to carry out a physical experiment (as distinguished from a thought experiment) that will bear on the phenomenon of interest. Sometimes they design the experiment to capture a *qualitative* effect, such as whether a certain phenomenon will occur under predicted circumstances. For example, will a current flow if we do such and such? In other situations, the scientist designs the experiment to capture a *quantitative* effect, such as how much of something will occur under particular circumstances. For example, how much current will flow if we do such and such?

A scientist who decides to carry out a physical experiment must decide:

1. Which experiment will be carried out (can the phenomenon be observed directly, or will indirect reasoning be necessary)?
2. Which equipment will be used to carry out the experiment? The equipment decision determines how large the observational errors are likely to be, and often, whether the phenomenon can be observed at all.
3. Which hypotheses will be tested by this experiment? The scientist must decide which conditions must be controlled or kept constant during the experiment and which can be varied to generate the phenomenon. The hypothesis about the predicted phenomenon must be clearly stated in advance of the experiment, and the conditions under which the phenomenon will occur must be prespecified and designed into the experiment. If the experiment is quantitative, the scientist must determine which outcome variables must be monitored when the input variables are varied.
4. How will the hypotheses be examined? That is, how "close" to the predicted phenomenon must the experimental outcome be before the scientist will claim to have produced the anticipated effect? How should the scientist measure closeness?
5. What extraneous variables should be controlled or measured?
6. If the scientist is to repeat the experiment on several occasions, how will strange or unusual outcome observations be treated? How many times should the experiment be replicated? How will the results of various replications be combined? How sensitive must the experiment be to observe the phenomenon of interest? What if the phenomenon were to be observed only sometimes and not at others?
7. What should be done if the experimental outcomes contradict the initial hypotheses?
8. How will the results be reported?
9. Which results will be reported to the scientific community so that the work can be reviewed by peers and replicated by others to verify its correctness?

All of these choices the scientist must make are partly subjective and partly objective. While attempting to avoid obvious biases, a scientist wants to use the fundamental understanding of the underlying phenomenon to help arrive at scientific conclusions that will be meaningful. To ignore such preexisting understanding of the phenomenon under study is to fail to take advantage of the best information available to build new knowledge. It would amount to reinventing the wheel every time we wanted to make an improvement in our understanding of transportation methods. We see in Chapter 5 how Bayesian methods of analysis seek to take advantage of this pre-existing understanding by incorporating various aspects of it into the analysis of an experiment.

1.7 ANALYSIS OF RESULTS OF PHYSICAL EXPERIMENTS

The results of physical experiments that scientists carry out are observations, either qualitative or quantitative. If qualitative, they may be *descriptive*, in that they explain in qualitative terms what happened; or they may be *categorical*, in that they summarize whether the predicted phenomenon occurred. When quantitative, they are numbers obtained by observing the phenomenon under study under various conditions. Any of these forms of outcome of an experiment characterizes the results of the experiment. These are the basic raw data. After obtaining such data, the scientist must try to make sense out of them. If the data are quantitative, they must be summarized and understood visually. In any case, inferences and predictions must be made from them. Quantitatively observed data points generally contain errors of measurement. Such errors of measurement are largely random, and thus follow the laws of probability, part of the domain of statistical science. Also in the domain of statistical science is the modeling of the laws that are thought to govern the phenomenon.

The scientists discussed in Chapter 4 did not use the methods of analysis of Bayesian statistical science (or for that matter, with the exception of Einstein, any formal statistical methods). Although it is true that Bayes' theorem had been put forth in the middle of the eighteenth century, many of the methods of analysis that today we call Bayesian statistical inference had not yet been developed; and none of it was available during the lifetimes of many of our scientists. Such methodology, however, is now available for scientific analysis, and makes possible more precise inference from the results of categorical and quantitative experiments. Such methodology is discussed at length in, Chapter 5.

1.8 BLINDING AGAINST EXPERIMENTER BIAS

The subjectivity of scientists has long been well-known to scientists themselves, and consequently, they have established procedures to guard against those effects of subjectivity that may bias experiments and mislead experimenters. In biology/medicine/pharmacology, for example, modern clinical statistical trials have typically required that allocation of treatments to subjects be decided on a randomized basis with some subjects receiving an inert placebo, so that causation can be established. But the safeguard against experimenter bias is via a process called *double-blinding*. In a double-blind experiment, neither the people who diagnose the results of the clinical trials nor the subjects themselves know which subjects actually received the drug being tested for efficacy and which received the placebo. If the experimenter does not know whether the subject he or she is diagnosing has received the treatment or the placebo, his or her diagnostic judgment cannot be influenced by that knowledge. Similarly, a patient cannot attribute changes in his or her condition to a treatment if he or she is not sure whether indeed he or she is getting the treatment or getting a placebo.

In the social sciences, the existence of "experimenter effects" [for example, Rosenthal and Jacobson (1968)] in which the person conducting an experiment is likely to find the results he or she expected, has given rise to a set of procedures by which the person actually running the experiment is kept blind to the hypothesis being tested.

In our discussion of Gregor Mendel in Section 3.3 we see that one of the reasons suggested for his data being "too good to be true" is that when one expects a certain ratio of results and can see that ratio developing as the data are counted, one may be unconsciously led to stop counting when the ratios are as expected. Present-day genetics researchers routinely blind themselves to the results of such counts as they are developing.

In his oil drop experiment to study the charge on the electron, Robert Millikan used a preferential treatment of his data. While he seemed to be choosing to eliminate only observations with large experimental error, in fact he eliminated observations that were contrary to his preconceived ideas, thus permitting his personal biases to determine his inferences about the true value of the charge on the electron. (See Section 3.4 for a discussion of how he weighted his data to conclude what he preferred.)

Recent efforts by "particle physicists" have resulted in a blinding approach to guarding against experimental bias in their field, an approach that has been called *offsets*. "Research teams in particle physics have been programming their computers to add unknown numbers called 'offsets,' to their data to make the outcome of their analyses blind . . . It is only after they experiment is over that the researchers discover what the value of the offset is" (Glanz, 2000, p. D1).

Dr. Eric Prebys, a particle physicist at Princeton University and a member of a large research team called Belle, said scientists "always want a particular outcome whatever they say. Removing all landmarks is the only sure way to insulate them from their own hopes and expectations" (quoted in Glanz, 2000, p. D4). The basic idea is to blind the researchers from seeing the final experimental results until after they have made their inferences about what phenomena actually generated those experimental results. The focus of the experiment in which Dr. Prebys has been involved is the small difference between the physics of matter and the physics of antimatter. Where are the asymmetries? Currently physicists have strong preconceived predilections to be able to confirm certain asymmetries in the physics of matter and antimatter, partly because such a finding would constitute scientific "news." Evidence so far has demonstrated some asymmetries; more are being sought to strengthen the theory. The current problem they have is trying to carry out experiments that relate to this issue without letting their personal biases about what to expect about symmetry/asymmetry creep in. As one part of the analysis Dr. Prebys is involved in, "they tried to determine whether a particular decay signature was due to a particle or its associated antiparticle" (Prebys, 2000). A *decay signature* is a term used by particle physicists to mean that the pattern of data that they have about a particular subnuclear particle uniquely identifies the type of particle that generated those data.

The blinding of the analysts was accomplished by using a computer subroutine that randomly returned a "1" for a particle or a "−1" for an antiparticle, effectively hiding the asymmetry that the scientists were looking for. The research team only examined the real experimental results as a final step after agreeing to accept the inferences about what had caused the results obtained, whatever those results might be. When they made their inference about what had caused the results that they had found, their inference implied that there was no asymmetry! When they reexamined the experimental results that had been hidden, they saw that that had they used only results with small measurement error, and ignored those with the larger measurement error (since such a preferential treatment of the data strongly supported their biases), they would have concluded that they had indeed found asymmetry in the physics of matter and antimatter. Because of the blinding effect of the offsets they were forced to conclude there was no asymmetry. It is certainly possible that future experimenters will conclude that there is in fact asymmetry, but that it was difficult to detect in Dr. Prebys' experiment.

It is not yet known how widespread this "offsets" or "blind" approach will become. Its function is to remove the possibly misleading aspects of a scientist's subjectivity, aspects that can bias him/her into misinterpretation of experimental results. It cannot (and should not) eliminate all the subjectivity of physical scientists. The physical scientist still must decide which experiment to carry out, he or she must still design the experiment, and must still decide which models and analyses will be used. Breakthroughs in science are likely to occur only with generous applications of informed understanding of the phenomena being studied, and with continued use of hunches and intuition. Physics is likely to benefit from the uses of offsets because such methodology added to the armamentarium of physics will result in an improved mixture of subjectivity and objectivity, the essence of all good science.

REFERENCES

Glanz, James (2000). "New Tactic in Physics: Hiding the Answer," *Science Times*, in *The New York Times*, August 8, 2000, pp. D1, and D4.

Hacking, Ian (1965). *Logic of Scientific Inference*. Cambridge: Cambridge University Press.

Kuhn, Thomas S. (1962). *The Structure of Scientific Revolutions*, 2nd ed. Chicago: University of Chicago Press.

Prebys, Eric (2000). Personal communication, August 23, 2000 and August 27, 2000.

Roberts, Royston M. (1989). *Serendipity: Accidental Discoveries in Science*. New York: Wiley.

Rosenthal, Robert, and Lenore Jacobson (1968). *Pygmalion in the Classroom: Teacher Expectation and Pupils' Intellectual Development*. New York: Holt, Rinehart and Winston.

Wolpert, Lewis (1992). *The Unnatural Nature of Science*. Boston: Harvard University Press.

CHAPTER 2

Selecting the Scientists

How have we decided which scientists to study? In Chapter 4 we study some of the most famous scientists in all history, and we examine the nature of the subjectivity and objectivity that they exercised in their approaches to their research. But *we* did not select which scientists we should study; we left that task to someone else.

The reader might guess that we carefully hand-picked scientists whom we knew to be considerably less than objective in their approaches to scientific research, just to make our point. But we did not exercise such subjectivity in Chapter 4, although we did in Chapter 3. In that introductory treatment of celebrated cases of extreme subjectivity we did not plan to be objective in our choices of scientists to study. For Chapter 3 we selected for study several scientists who have been subjects of notoriety, not only because their scientific achievements have made them so renowned, but also because of their now well-known uses of subjectivity in their work. The five scientists examined in Chapter 3—Johannes Kepler, Gregor Mendel, Robert Millikan, Cyril Burt, and Margaret Mead—are discussed but briefly, merely enough to review some of the major subjective aspects of some of their research.

But for the largest chapter of this book, that is Chapter 4, we attempted to approach this important selection question much more objectively ourselves. In that chapter we examine the cases of 12 world-famous scientists in considerable detail: some from ancient times, some from the Renaissance and the period of the scientific revolution, and some from the modern era. But we did not select these 12 scientists. To be as objective as we could in our selection of the scientists we would use to demonstrate our contention, we decided to work from a template not selected by us. We accomplished this by focusing on someone else's book in which the lives of some eminent scientists were already discussed, and then by our doing a case study of every one of those scientists preselected for us by the author of that book. The book we chose is *The Great Scientists*, by Jack Meadows (1987). Meadows describes the lives, times, and scientific achievements of the following 12 scientists:

1. Aristotle
2. Galileo Galilei
3. William Harvey
4. Isaac Newton
5. Antoine Lavoisier
6. Alexander von Humboldt
7. Michael Faraday
8. Charles Darwin
9. Louis Pasteur
10. Sigmund Freud
11. Marie Curie
12. Albert Einstein

These are therefore the cases we examine in Chapter 4. There, in each case, we provide a review of the life of the scientist, a review of his or her scientific contributions, and an examination of the ways this scientist used subjectivity in his or her research.

Why did we select the Meadows book, in particular, to dictate our cases? Should we be concerned about whether these scientists are universally agreed upon as the most famous, most important, or even most appropriate scientists to study? In both Chapters 3 and 4 we try, as much as possible, to look at what the scientists actually did, rather than what they say they did or what others say they did. In that spirit, we must note that we chose the Meadows book as the basis for our list of scientists partly because it seemed to satisfy certain intuitive criteria of comprehensiveness of coverage and importance of the scientists included, but otherwise, we chose it somewhat arbitrarily. As we discussed our project of demonstrating the ways in which scientists used personal beliefs, intuition, and previous knowledge, and pondered how to make that demonstration convincing to others as well as to ourselves, we realized that we would need to show that we had subjected our own ideas to the kind of objective test that would be acceptable only if we relied on an outside agent to choose the cases on which our argument would rest. The Meadows book was at hand, and we decided to accept the challenge of demonstrating the subjectivity in the work of the 12 scientists that Meadows had designated as "great."

In retrospect, despite the somewhat arbitrary method by which we chose it, we believe that our choice of the Meadows book was a good one. These are our reasons for that judgment:

1. It purported to treat some of the most famous scientists in history.
2. It treated as many as a dozen scientists, while some other books of similar genre treated perhaps only half as many, or fewer.
3. It was not a dictionary or encyclopedia of scientists, which would be too unwieldy to handle.

4. It was a recent book.
5. It treated scientists of many nationalities and time periods.
6. It included scientists from many scientific fields, including biology, chemistry, geography, mathematics, medicine, psychology, and physics.
7. It treated all scientists equally, in that it devoted precisely 20 pages, including pictures, to each scientist.

The scientists selected by Meadows surely include some of the most famous; but it is difficult to rank scientists. What criteria should be used: Some measure of how well known they are? Importance of their accomplishments? (How does one measure importance?) How often other scientists cite their work? How many of their ideas were eventually converted into applications used by society at large? Or which of countless other criteria should be used? People who do such rankings differ among themselves on how scientists should be ranked. For this reason we have examined a recent book of Simmons (1996) of rankings of the top 100 scientists (in his admittedly subjective view) and have included his rankings of the Meadow's scientists in the accompanying table, "Coverage of Scientists in Selected Reference Books." We have also examined several other books (see the References at the end of the chapter) that purport to include the most famous scientists. We have recorded in the table the extent to which these books have selected the Meadows scientists to examine in terms of the numbers of pages each book has devoted to each of the Meadows scientists. Although the list of comparison books is surely not exhaustive, it does reflect the results of a fairly extensive library search.

The table shows, for example, that while Charles Darwin was ranked fourth among the top 100 scientists by Simmons and had the obligatory 20 pages devoted to him by Meadows, the Wilson book devoted 28 pages to him, whereas the Bolton and Posin books did not discuss him at all. Scanning across the table shows that in this sampling of books devoted to biographical accounts of these great scientists, the authors varied greatly in their decisions as to how to weight the amount of coverage they would devote to each case. Examining the table further shows us that many of the Meadows scientists are included among the top scientists in the Simmons rankings: Some are near the top, some are quite a bit farther down: and two are not even included among the top 100. (Simmons notes that Aristotle is of the greatest importance in the history of science but excludes him because his contributions are not due to direct influence; he excludes Alexander von Humboldt without any comment.) But in Chapter 4 we treat every one of the Meadows scientists as one of our cases, to be as objective and even-handed as possible. The variation in number of pages devoted to each scientist by the various authors, as reflected in the table, illustrates, at least in part, the differing views the authors have about rankings of these scientists. However, we can also note from the table that across 12 books purporting to treat the world's most famous scientists, there is substantial agreement as to who the most famous scientists were.

Coverage of Scientists in Selected Reference Books

Scientist	Rank in Simmons (1996)	Asimov (1989)	Chambers (1989)	Crowther (1995)	Simmons (1996)	Bolton (1960)	Hamilton (1991)	Wilson (1937)	Lenard (1933)	Posin (1961)	Science Digest (1954)	Dolin (1960)	Gibson (1970)
						Number of Pages in ….							
Scientists Discussed in Meadows (1987)													
Newton	1	19	2.5	47	4	14	6	19	28	20	4	4	27
Einstein	2	17	2.5	46	6	15	8	24	0	12	3	3	0
Darwin	4	8	1.5	43	5	0	6	28	5	0	4	10	13
Pasteur	5	11	2	0	5	0	6	31	0	0	6	10	0
Freud	6	6	0	0	6	0	0	0	0	0	5	0	0
Galileo	7	20	1	39	5	23	0	32	15	12	6	8	31
Lavoisier	8	9	1.5	0	4	0	6	12	0	0	3	9	0
Faraday	11	10	2	0	5	7	8	26	16	20	6	7	17
Curie	26	5	1.5	0	4	13	0	0	0	10	4	9	0
Harvey	38	2	0.5	0	4	0	0	10	0	0	5	6	0
Aristotle	—	9	0.5	0	0	0	0	14	1	0	0	0	5
von Humbolt	—	0	1	0	0	0	0	0	0	0	0	0	0
Scientists Not Discussed in Meadows (1987)													
Kepler	9	0	0.5	0	4	0	0	28	9	12	0	4	9
Mendel	60	0	1	0	3	0	6	12	0	0	6	0	0
Mead	—	0	0	0	—	0	0	0	0	0	0	0	0
Millikan	—	0	0.5	0	—	0	0	0	0	0	0	0	0
Burt	—	0	0	0	—	0	0	0	0	0	0	0	0

It will be seen that in every case, subjectivity played an important role in the research of the scientists we studied. Since this is how these most famous people approached science, it is not unreasonable to expect that most scientists probably have approached their own research in analogous fashion.

REFERENCES

Asimov, Isaac (1989). *Asimov's Chronology of Science and Discovery*. New York: Harper & Row.

Bolton, Sarah Knowles (1960). *Famous Men of Science*, revised by Barbara Lovett Cline. New York: Thomas Y. Crowell.

Chambers Concise Dictionary of Scientists (1989). Cambridge: W & R Chambers and the Press Syndicate of the University of Cambridge.

Crowther, J. G. (1936). *Men of Science*. New York: W.W. Norton.

———(1995). *Six Great Scientists: Copernicus, Galileo, Newton, Darwin, Marie Curie, Einstein*. New York: Barnes & Noble.

Dolin, Arnold (1960). *Great Men of Science*. New York: Hart.

Gibson, Charles (1970). *Heroes of the Scientific World*. Freeport, NY: Books for Libraries Press.

Hamilton, John (1991). *They Made Our World: Five Centuries of Great Scientists and Inventors*. London: Broadside Books.

Lenard, Philipp Eduard Anton (1933). *Great Men of Science: A History of Scientific Progress*. New York: Macmillan.

Meadows, Jack (1987). *The Great Scientists*. New York: Oxford University Press.

Posin, Dan Q. (1961). *Dr. Posin's Giants: Men of Science*. Evanston, IL: Row, Peterson.

Science Digest (1954). *Science Milestones: The Story of the Epic Scientific Achievements and the Men Who Made Them Possible*. Chicago: Windsor Press.

Simmons, John (1996). *The Scientific 100: A Ranking of the Most Influential Scientists, Past and Present*. Secaucus, NJ: Carol Publishing Group, Citadel Press.

Wilson, Grove (1937). *Great Men of Science*. Garden City, NY: Garden City Publishing Company.

CHAPTER 3

Some Well-Known Stories of Extreme Subjectivity

3.1 INTRODUCTION

In this chapter we examine the work of five scientists who have become very celebrated cases of the excessive use of subjective methods in science. They are all very famous, highly respected scientists, and deservedly so. They are icons of science. These famous scientists all took advantage of their intuition, hunches, beliefs, and informed understanding of the processes they were studying and combined them informally with experimental data to arrive at scientific conclusions. But they are grouped here and studied because it is important to understand that even famous scientists may sometimes have overstepped the bounds of what is considered acceptable scientific practice in their zeal to rush to scientific conclusions and to have their work accepted by their peers. Although we advocate the combining of informed judgment with experimental data and show in Chapter 5 how to do this formally using Bayes' theorem, we do not condone the fabrication of data, the unjustified manipulation of data to select only those data that support a preconceived hypothesis, or any other fraudulent excessive use of subjectivity.

3.2 JOHANNES KEPLER (1571–1630)

Johannes Kepler was a German astronomer–mathematician who discovered from both empirical data and informed theorizing that the orbits in which the earth and the other planets travel around the sun are elliptical. He put forth the three fundamental laws of planetary motion in use today and is considered the father of modern astronomy. The Keplerian laws were later used by Sir Isaac Newton to develop his own laws of gravitational force of attraction of heavenly bodies (published in Newton's book, *The Principia*, 1687). Kepler's laws are:

1. All the planets in our solar system move around the sun in elliptical orbits, having the sun as one of the foci (1609).

2. A radius vector joining any planet to the sun sweeps out equal areas in equal periods of time (1609).

3. The squares of the periods of revolution of the planets are directly proportional to the cubes of their mean distances from the sun (1618).

We discuss below briefly how the discovery of these laws came about, and how Kepler made generous use of his own personal beliefs in his methodological development and in his communication of the laws to the scientific community and the rest of the world.

The story of Johannes Kepler really begins with Ptolemy of Alexandria, Egypt, a scientist who, in about A.D. 140, postulated the geocentric view that the earth is at the center of the solar system and that all the planets and the sun revolve around it. This view of the world was the one that prevailed in the Western world for about the next 1300 years of civilization, until Nicholas Copernicus (1473–1543). Copernicus was a Renaissance astronomer from East Prussia, now Poland. He described the solar system in heliocentric terms, with the sun, fixed in place, at the center, and Earth and all the other planets (six were known at the time) revolving around it in circles. Copernicus, by rejecting Ptolemy's geocentric theory and adopting the heliocentric view, got it basically right, although he believed the planetary orbits to be circles rather than ellipses, as they actually are, and believed that the sun is fixed in space.

Kepler's other major precursor, Tycho Brahe (1546–1601), was a Danish astronomer born about the time that Copernicus died. Brahe devoted his life, first from Denmark and then from Prague, to observing accurately with his naked eye, and recording the positions and motions of the heavenly bodies. His observations produced a comprehensive study of the solar system and the recording of accurate positions of more than 777 "fixed stars," a scientific treasure trove of accurate observational data on the stars and planets. Brahe rose to become the imperial mathematician of the Holy Roman Empire.

At college, Kepler studied astronomy under Michael Mastlin, a professor who was unusual in that he was one of the few who was convinced by the work of Copernicus that the solar system was heliocentric. Kepler was convinced in turn. Kepler sent Brahe a copy of his first major work in astronomy (*Cosmographic Mystery*, written in Latin), work that built on Copernican theory. Brahe was so impressed by both the astronomy and the mathematics that although he could not agree with the Copernican theory, he invited the 24-year-old Kepler, in 1600, to join his laboratory just outside Prague.

Brahe asked Kepler to work out the orbit of Mars. When Brahe died a year later, Kepler became keeper of Brahe's repository of observational data; he was also appointed Brahe's successor as imperial mathematician for the Holy Roman Empire. As such, he not only provided the emperor with some 800 astrological horoscopes, but also acquired a team of astronomers to assist him.

Kepler continued to work on the orbit of Mars. He tried unsuccessfully to fit Brahe's data on Mars to a circle, to show that Mars moved in a circle around the sun, but he failed. He finally realized the truth, that it must move in an ellipse (he also tried an oval and other geometric shapes). He then generalized the concept to all the planets. With the help of Tycho Brahe's database, Kepler became determined to quantify the motions of all the planets, and in 1609 he published a 70-chapter tome, *Astronomia Nova*, a volume that summarized not only his work on the orbit of Mars but also two of his famous three laws of planetary motion.

Almost 400 years later, William H. Donahue undertook the task of translating Kepler's 1609 *Astronomia Nova* into the English *New Astronomy* (Donahue, 1992). When, in the course of this work, he redid many of Kepler's calculations, he was startled to find some fundamental inconsistencies with Kepler's reporting of the same calculations (Donahue, 1988). Writing of Donahue's pathbreaking work in the *New York Times*, William Broad (1990) summarized Donahue's findings as saying that although Kepler claimed to have confirmed the elliptical orbits by independent observations and calculations of the position of Mars, in fact Kepler derived the data from the theory instead of the other way around.

In his 1988 article, Donahue had noted that Kepler is reputed to have plotted the orbit of Mars by triangulation and then seen that the plot formed an ellipse: that is, the positions in the sky of Earth, the Sun, and Mars form a triangle, at any given point in time. At a later time, they form a new triangle. To understand the paths of motion involved, Kepler needed to measure accurately the vertices of such triangles and the time elapsed between one triangular position and another. But a close study of Kepler's *New Astronomy* (1609) shows that the plotted points do not fall exactly on such an ellipse. (Of course, measurements rarely fall exactly on a theoretical curve because they usually have random error sources incorporated into them.) Curtis Wilson (1968), however, carries the error argument further. The lack of precision inherent in the method of triangulation would have *forced* Kepler to use the plotted points only as a guide to his theorizing. Donahue tells us that Kepler presents a large table that supposedly shows the results of such observations, but the numbers in the table do not agree with the computations from which Kepler claims they were made.

After detailed computational arguments, Donahue concludes that the results reported by Kepler in the table were not at all based on Brahe's observational data; rather, they were fabricated on the basis of Kepler's determination that the Mars orbit was elliptical. Donahue reasons that Kepler must have gone back to revise his earlier calculations that were made prior to his understanding that the orbit of Mars was actually elliptical. Thus anyone who cared to check Kepler's tables would find numbers that are consistent with the elliptical orbit postulated for Mars and would be inclined to believe that the numbers represented observational data. In fact, they were computed from the hypothesis of an elliptical orbit and then modified for measurement error.

Such data, if they were truly observations, would be prima facie evidence of the theory's correctness.

So Donahue concludes (1988, p. 236) with the following understanding of Kepler's thinking when he realized that the theory was not obviously derivable from the observations and chose to present calculations that did not match the original data: "Not only would the numbers be confused, but Kepler saw clearly that no satisfactory theory could come from such a procedure. So rather than obliterate the traces of his earlier struggles, or present them in their messy detail, he chose a short cut." What appears to have occurred here is that a great scientist developed enormous understanding of the underlying processes that he was studying. That understanding was what he needed to develop the three planetary laws for which he became so famous. But later, he seems to have crossed the line of acceptable scientific practice. He became so convinced of what drove these physical processes that he subjectively projected his personal, nonobservationally-based beliefs onto the reporting scene to convince others in the scientific community of the validity of his theories. Fortunately, he was brilliant enough to get it right. This is a common thread that has run through the methodology of many of the world's successful scientists.

REFERENCES

Broad, William J. (1990). "After 400 Years, a Challenge to Kepler: He Fabricated His Data, Scholar Says," *Science Times in the New York Times*, January 23, pp. B5, B7.

Donahue, W. H. (1988). "Kepler's Fabricated Figures: Covering Up the Mess in *The New Astronomy*," *Journal for the History of Astronomy*, **19**:216–237.

————(Trans.) (1992). *The New Astronomy* by Johannes Kepler. Cambridge: Cambridge University Press.

Encyclopaedia Britannica (1975). Vol. V, p. 766.

Wilson, Curtis (1968). "Kepler's Derivation of the Elliptical Path," *Isis*, **59**:75–90.

3.3   GREGOR MENDEL (1824–1884)

Gregor Mendel was the founder of modern genetics. He was born in Moravia, now in the Czech Republic. He attended courses in philosophy at Olmutz University for two years and then joined the Augustinian monastery at Brün as a means of attaining further education. He was ordained in 1848. He was sent by his order to the University of Vienna for two years to study natural science, physics, and mathematics, and to prepare for an examination to become a regular high school teacher.

During his time in Vienna he was exposed to botanists studying issues of plant evolution and was a demonstrator in the Physical Institute. Here, Olby

(1966, p. 113) points out: "He would have had ample opportunity for becoming familiar with the physicist's approach to experimentation which, unlike that of the naturalist, is not to make observations in a Baconian manner and then seek for an underlying pattern but to analyse a problem first, arrive at a solution on paper, and only then to carry out an experiment to confirm or refute the solution. With this approach in mind, the scientist attempts to design an experiment that will bear meaningfully on the hypothesized pattern. Of course, the problem is suggested in the first place by observational data. . . ." At Vienna, too, Mendel would have been exposed to the facts then known about hybrids—that there is uniformity in the F_1 generation and reversion in the F_2 generation. (The F_1 generation are the offspring of a hybridization where two plants that are genetically pure have been cross-fertilized; the F_2 generation are the offspring by self-pollination in the F_1 generation.)

In 1854, Mendel returned to substitute teaching at the secondary level, and in 1855 he again failed the examination for regular teacher and so remained a substitute teacher of physics and natural history for 14 years. It was during this period that Mendel carried out his experiments in plant hybridization. The work on peas, on which Mendel's fame rests, started in 1856 and culminated with a paper that was read before the Brünn Society for Natural History in 1865 and published in their proceedings in 1866. The ideas that Mendel presented in that paper have become the basic principles of genetics: Heredity particles (now called genes) are passed from parent to offspring, where they remain intact, neither influencing each other nor changing. These genes determine the characteristics of the offspring of the next and of further generations according to a set of simple statistical rules. These ideas provided a mechanism through which the evolution of species by natural selection theorized by Darwin could proceed (see the discussion of Darwin in Section 4.9).

To understand Mendel's ideas, let us take as an example human eye color. (For the sake of clarity, we shall use modern terminology and even concepts that were unknown to Mendel.) Every human being has 46 chromosomes (23 pairs) in every cell (except the germ cells, that is, the eggs and sperm, which have only 23 chromosomes each, one from each pair). On each chromosome there are many particles called genes. The genes on each pair of chromosomes are themselves paired, so that in the case of eye color, a brown-eyed male must have a gene for brown eyes on one chromosome, and on the other chromosome he must have either another gene for brown eyes, or a gene for a different color, say blue. Because brown is dominant, the combination of one blue-eyed gene and one brown-eyed gene results in a person with brown eyes. For purposes of this explanation, let us assume that the man's other eye-color gene is the recessive gene for blue eyes. This means that when his germ cells are formed, each with one of the pair of chromosomes that carry the gene for eye color, half will have the gene for brown eyes and half will have the gene for blue eyes.

Let us examine the possible results of a mating of this brown-eyed man with a brown-eyed woman; for the sake of this explanation, we shall assume that she also carries both the gene for blue eyes and that for brown eyes (and hence half of her germ cells carry the gene for blue eye color and half for brown eye color). We find that there are four possible genetic outcomes for eye color in the child: genes for blue eye color from both parents, genes for brown eye color from both parents, a gene for blue eyes from the mother and a gene for brown eyes from the father, and a gene for brown eyes from the mother and a gene for blue eyes from the father. But the last two possibilities result in the same brown eye color as occurs with two brown genes, because brown is dominant. So the four genetic possibilities (genotypes) reduce to two eye-color appearances (phenotypes).

Mendel's great insight was to predict the likelihood that each genotype would occur. In the case of our eye-color example, he would say that the blue–blue combination and the brown–brown combination are equally likely, but the blue–brown combination occurs twice as often because we might get a blue-from-mother-brown-from-father, or a brown-from-mother-blue-from-father. An equivalent interpretation is that there is a 25% chance that the child would have blue–blue genes for eye color (called a *double recessive*), a 25% chance that the child would have brown–brown genes for eye color (called a *double dominant*), and a 50% chance that the child would have the brown–blue gene combination (called a *hybrid*). These percentages are really probabilities. They mean that if the mating couple were to have eight children, it would be possible for all eight to be blue-eyed, but it is extremely unlikely for that to happen, just as a well-balanced coin flipped randomly eight times could result in eight heads, but "don't bet on it" (the chance of eight heads in eight flips is 1 in 256). The couple would be most likely to have two children with blue–blue genes, two with brown–brown genes, and four with blue–brown genes—but even that outcome is only an on-the-average phenomenon.

If we now translate these ideas about eye color to Mendel's experiments, the same principles apply. Mendel worked with hybrid peas, that is, peas that he had bred from pure strains. He thus knew the hybrids to have one gene for the dominant variant of the characteristic and the other gene for the recessive variant of the characteristic. (Mendel worked with seven characteristics in all, having to do with such things as where the flowers of the pea plant were located on the plant, the shape of the seed, the color of the seed, and so on.) Then, when he self-fertilized such a hybrid pea plant, the offspring should have occurred in the same genetic ratios as in the case of blue and brown eye-color genes: that is, 25%, 50%, and 25%.

Let us use as a specific example the sixth of the seven characteristics of peas that Mendel used—the position of the flowers on the stem. He found a variety of peas that consistently distributed its flowers along the main stem (termed *axial*) and another that consistently bunched the flowers at the top of the stem (*terminal*). When these two varieties were used in a hybridization experiment, each member of the next generation (the F_1 generation) would have in each cell one gene that comes from the parent with axial flowers and one from the parent with terminal flowers. Half the reproductive cells of each offspring would carry the gene for axial flowers and half for terminal flowers. As axial flowers are dominant, all offspring in the F_1 generation have axial flowers.

Mendel then self-fertilized members of the F_1 generation. (Self-fertilization assured that Mendel knew that the source of both germ cells going to form the next generation was the same, and thus that method of fertilization was used throughout his experiments on plant hybridization.) In the next generation, the F_2 generation, he found that three-fourths of the offspring had axial flowers and one-fourth had terminal flowers. He reasoned that those with terminal flowers had received the recessive gene from both reproductive cells, but needed to breed another generation of hybrids via self-fertilization to distinguish between those plants with axial flowers that carried two dominant genes and those that carried one dominant (for axial flowers) and one recessive (for terminal flowers). Those plants in the F_2 generation that produced only offspring with axial flowers would be considered to carry two dominant genes; those in the F_2 generation that produced both axially flowered and terminally flowered offspring in the F_3 generation would necessarily be carrying one dominant and one recessive gene. Mendel writes that he chose 100 members of the axially flowered plants in the F_2 generation and from each planted 10 seeds to make this determination. He then reports that in the F_3 generation, the offspring of 33 plants had only axial flowers, and thus the parents were classified as carrying two dominant genes and that the offspring of 67 plants had some plants with axial flowers and some with terminal flowers, so the plants of the F_2 generation were classified as hybrids.

The experimental activities largely ceased when Mendel was elected abbot of his monastery in 1868 and his time was taken up with administrative duties and later by a controversy with the government about taxes on monastery property. Mendel continued, however, to act as a man of science and to make meteorological and horticultural observations until his death on January 6, 1884. Mendel's work was practically unknown for 35 years, despite wide circulation of the journal in which it appeared and despite his correspondence with Carl von Nägeli, then the foremost expert on hybridization. Its chief principles were then discovered independently and approximately simultaneously by three botanists (Correns, 1900; de Vries, 1900; Von Tschermak, 1900). By that time, according to Mangelsdorf (see Mendel Centennial Symposium, 1967) "in large part because of the profound observations of the German zoologist Weismann, father of the germ plasm theory, the scientific world was ready and the principles of inheritance had an immediate impact, almost explosive in its nature, upon biological thinking and research."

There has been a controversy about Mendel's use of subjectivity in his work since R. A. Fisher, a prominent geneticist and often considered the founder of the modern field of statistical science, pointed out (Fisher, 1936) that Mendel's data seemed "too good to be true." Mendel's reports of the results of his matings of hybrid pea plants, embodied in the ratio of double recessive, double dominant, and hybrid offspring, approximated extremely closely the theoretical percentages of 25%, 25%, and 50%. But we just don't expect such good agreement in so few matings, just as we don't expect the well-balanced coin to fall heads exactly 5 times every time we flip it 10 times. Indeed, Fisher found, using a form of statistical test called *chi-square*, that Mendel's results were so close to what would be expected that such agreement could happen by chance less than once in 10,000 times (1936, p. 131).

But Fisher also pointed out an even more serious flaw in Mendel's work. This had to do with how Mendel would decide whether an F_2 generation pea plant was a hybrid or a dominant–dominant plant. As we have seen, part of his task in the experiments was to classify the plants in the F_2 generation that showed the dominant phenotype as dominant–dominant or dominant–recessive. If the characteristic Mendel was working with concerned the plant's seeds (e.g., whether they were "round" or "wrinkled"), he merely had to examine the seeds—a dominant–dominant plant would bear only seeds showing the dominant characteristic, while a hybrid would bear some seeds showing the dominant phenotype and some showing the recessive. But if the characteristic had to do with the conformation of the plant itself (e.g., the placement of the flowers on the stem), the dominant–dominant and hybrid plants looked the same and Mendel couldn't be sure which was which. It was necessary to breed another generation, the F_3 generation, through self-fertilization of the plants to be classified and to examine whether the double recessive characteristic appeared in any of the offspring. If it did not, the parent plant was characterized as having the dominant–dominant gene combination. But misclassifications could easily occur by such methods in the case where only a few offspring were examined and by chance all the examined offspring of a hybrid happened to be double dominant. Fisher did the calculations and found that when Mendel used 10 plants in the F_3 generation for each parent to be classified, about 5.6% of the hybrids would be misclassified as dominant–dominants. Therefore, the ratio of plants classified as carrying two dominant genes to those carrying one dominant and one recessive among those in the F_2 generation which displayed the dominant axial flowers should have been expected to be not $1:2$ but $1.1126:1.8874$. Mendel's data, however, fit the $1:2$ ratio very well, and fail to fit the adjusted ratio by a margin that would be expected to occur less than once in 2000 trials.

Another prominent geneticist, Sewell Wright (1966), repeated Fisher's analysis a generation later and achieved essentially the same results as Fisher had. Various explanations have been advanced for Mendel's "too-good" data. All hinge on the idea that Mendel had formulated his theory before collecting his data. Whereas this is common practice for scientists in all fields of study today and is indeed considered part of the "scientific method," it was unusual for naturalists in the mid-nineteenth century. They worked habitually in the style advocated by Francis Bacon, amassing great quantities of data and then spending years searching for patterns therein. Mendel undoubtedly absorbed the methods of experimental design from his work with physicists at the University of Vienna. The argument goes that Mendel either designed his experiments as demonstrations of the theory and discarded ill-fitting data as erroneous, or (whether consciously or unconsciously) biased the data collection efforts. That is, Mendel's subjective belief about the predictions of his theory about inheritance in successive generations may have been so strong that he doctored the data to fit the theory, either consciously or unconsciously.

Fisher (1936) himself is a spokesman for the view that Mendel's "experiments" were more demonstration than exploration. Indeed, Fisher pointed out that the conditions of knowledge by the middle of the nineteenth century were such that someone had only to assume that the material of inheritance was particulate rather than infinitely divisible and contributed by both parents to deduce the laws of inheritance. Fisher marshals evidence that the theory guided the design of the experiments by pointing out that Mendel was able to find seven pairs of varieties of peas, each differing in a single factor, perhaps choosing the factors to be studied as he chose the varieties; varieties that differed on more than a single factor would have introduced complications that would have made the demonstration of the theory more difficult. Fisher goes on to write (1936, p. 133): "Next, it appears that Mendel regarded the numerical frequency ratios, in which the laws of inheritance expressed themselves, simply as a ready method of demonstrating the truth of his factorial system, and that he was never much concerned to demonstrate either their exactitude or their consistency."

How the data came to agree too closely with expectations is also a subject of speculation. Fisher suggests that an assistant, knowing all too well what to expect, reported the data erroneously to Mendel. Others (e.g., Sturtevant, 1965) suggest that families which seemed aberrant might have been discarded as caused by experimental error, or that unconscious bias might have led Mendel or an assistant to classify doubtful individuals in a way that fit the expectations of the theory. Dunn (1965) gives a detailed explanation of another way in which such bias could occur. When tallying the phenotypes of offspring, especially into a large number of categories that are expected to be approximately equal, the investigator can watch the counts grow. Such an investigator may unconsciously be biased to stop the counting at a point when a large number of observations have been accumulated *and* the counts in the categories are indeed approximately equal. In the modern laboratory the dangers of such unconscious biases are recognized and protected against by setting up mechanisms to keep the investigator ignorant of the counts as they develop (A discussion of such safeguards against experimental bias in several scientific fields appears in Section 1.8.). But Mendel was a pioneer and had no such safeguards in place. But Dunn concludes (1965, p. 194): "There is no evidence of conscious fraud and he was careful to report wide deviations in some parts of some experiments which he would not have done if bent on fraud."

This absolution of Mendel from accusations of fraud is universal. Indeed, some authors, citing the care that Mendel exhibits in all his writings, suggest that there is no phenomenon of "too good" data that needs explanation. Glass

(1963, cited in Monaghan and Corcos, 1985b, p. 50) states this position strongly: "[It] was the existence of a tantalizing and hitherto insoluble problem that led him to undertake the particular kind of experiment he designed, and not the existence in his mind of any conceptual construct to which he was committed and for which he might have desired support."

Thus, as in the cases of other famous scientists who may have doctored or fabricated data (Cyril Burt, Johannes Kepler, Sir Isaac Newton), there are both supporters and detractors of Mendel. Several authors suggest that more than 10 plants per parent might have been grown in the F_3 generation, making the necessary correction to the expected ratio that Fisher pointed out less drastic. Others, for example, Weiling (1971) argue that other experimenters have found ratios that agree with theory at least as well as Mendel's do, and thus there may be something in the botanical process that invalidates the statistical procedures used by Fisher. Weiling suggests that because there is dependency between the fertilized cells of a single plant, the estimate of variance of the proportion of dominants would be smaller than that assumed by Fisher using a binomial distribution. If that variance, used in the denominator of the chi-square statistic, were replaced with something smaller, the computed value of the statistic would be larger and it might no longer point to congruence with expectations too good to be true. Citing Mendel's paper for some evidence of experiments that were repeated because they did not give the expected results, Van der Waerden (1968) suggests that Mendel used a system of sequential experimentation that would render Fisher's test invalid because that test is designed for an experiment that is fully planned in advance. Monaghan and Corcos (1985a) concur with an image of Mendel going back and forth between theory and data.

We shall not be able here to settle the question of whether Mendel's data have been doctored—indeed, the question may never be answered fully. But as will be noted from the discussion above, there are those who believe that the data are fabricated at least in part and those who believe that they are accurate reports of experiments as carried out. Both groups see Mendel's procedure as an interplay between theory and experimentation. They differ, primarily, on whether the theory was fully formed before the experiments began or was formulated piecemeal as results became available and was used to plan further rounds of experimentation. We must deplore any scientific fraud, even when carried out by a great scientist in the service of a correct theory; hence any fabrication of data by Mendel is indefensible. The interplay between theory and experiment, however, can be considered a form of subjectivity, employed to great profit by Gregor Mendel. Indeed, in his foreword to the centennial publication of Mendel's paper, Mangelsdorf (see Mendel Centennial Symposium, 1967) writes that "Mendel's work is a splendid example of what Henry Wallace has called 'small gardens and big ideas.' It is also one of the best examples of the preeminence of the human mind as a scientific instrument." Thus, Mendel certainly had the insight to develop a theory of heredity based on genes and chromosomes that has widespread currency today. Whether, like most scientists, he used his informed beliefs, intuition, and understanding of natural processes to aid him in arriving at meaningful scientific conclusions, or whether he sometimes fudged his data to make it conform with his preconceived theories, remains unclear.

REFERENCES

Beadle, G. (1967). "Mendelism," pp. 335–350 in A. Brink (Ed.), *Heritage from Mendel.* Madison, WI: University of Wisconsin Press.

Correns, Carl G. (1900). "Mendel's Regel über das Verhalten der Nachkommenschaft der Rassenbastarde," *Botanischen Gesesellschaft Berichte,* **18**:158–168.

Darbishire, A. D. (1911). *Breeding and the Mendelian Discovery.* London: Cassell and Company.

De Beer, G. (1964). "Mendel, Darwin, and Fisher," *Notes and Records of the Royal Society, London,* **19**(2):192–226.

de Vries, Hugo (1900). "Sur la Loi de Disjonction des Hybrides," *Comptes Rendus de l'Académie des Sciences, Paris,* **130**:845–847.

Dunn, L. C. (1965). "Mendel: His Work and His Place in History," *Proceedings of the American Philosophical Society,* **109**:189–198.

Fisher, R. A. (1936). "Has Mendel's Work Been Rediscovered?" *Annals of Science,* **1**:115–137.

Glass, B. (1963). "The Establishment of Modern Genetical Theory as an Example of the Interaction of Different Models, Techniques, and Inferences," pp. 521–541 in A. C. Crombie (Ed.), *Scientific Change.* New York: Basic Books; London: Heinemann.

Mendel Centennial Symposium (1965: Fort Collins (Colo.) (1967). *Heritage from Mendel: Proceedings of the Mendel Centennial Symposium Sponsored by the Genetics Society of America.* Madison, WI: Wisconsin University Press.

Mendel, Gregor (1965 [1866]). *Experiments in Plant Hybridisation.* Cambridge, MA: Harvard University Press.

Monaghan, F., and A. Corcos (1985a). "Chi-Square and Mendel's Experiments: Where's the Bias?" *Journal of Heredity,* **76**:307–309.

———(1985b). "Mendel, the Empiricist," *Journal of Heredity,* **76**:49–54.

Olby, Robert C. (1966). *Origins of Mendelism.* New York: Schocken Books.

Orel, V. (1968). "Will the Story on 'Too Good' Results of Mendel's Data Continue?" *Bioscience,* **18**:776–778.

Sturtevant, A. H. (1965). *A History of Genetics.* Tokyo: Weatherhill.

Van Der Waerden, B. L. (1968). "Mendel's Experiments," *Centaurus,* **12**:275–288.

Von Tschermak, E. (1900). "Uber künstliche Kreuaung bei *Pisum sativum,*" *Botanischen Gesellschaft Berichte,* **18**:232–239.

Weiling, F. (1971). "Mendel's Too Good Data," in *Special Issue on Pisum Experiments, Folia Mendelania,* **6**:71–77.

Wright, S. (1966). "Mendel's Ratios," in C. Stern and E. R. Sherwood (Eds.), *The Origin of Genetics: A Mendel Source Book.* San Francisco: W.H. Freeman.

3.4   ROBERT A. MILLIKAN (1868–1953)

Robert A. Millikan was an American physicist. He successfully measured the electric charge on a single electron, winning a Nobel prize in 1923 for this famous oil-drop experiment. He also won many other prizes and honors and a lasting place in the annals of science for his pioneering efforts. In 1909, Millikan began experimenting with both water and oil droplets, trying to measure the size of the charge on a single electron, under the assumption that the electron is a discrete particle; his early results were reported in Millikan (1909–1910). These experiments continued until about 1913, yielding results from increasingly refined procedures.

Millikan's basic experiment was to direct some oil droplets through a small hole in the top of a box with transparent sides, but with a metal top and bottom (these constituted condenser plates). He put a potential difference (voltage) across the top and bottom of the box and thereby created an electric field inside the box. A specific droplet of oil could be observed by a microscope and its weight calculated by the speed at which the droplet fell under gravity and against air resistance. Millikan was able to adjust the strength of the upward pull of the electric field on the oil droplet until the drop was just suspended motionless in the air, with the upward force on the charged oil droplet just equaling the downward pull of gravity. Millikan knew that the potential difference across the top and bottom of the box was just equal to the force of the electric field times the distance between the top and bottom plates of the box. Thus, knowing the distance and the voltage, he could readily calculate the force on the droplet, and, in turn, the charge on the electron (or electrons) in the droplet. As it turns out from his laboratory notebooks, he seems to have repeated the experiment 39 times but published only 29 of the sets of observations, discarding the rest (this is discussed further below). Millikan found that although his measurements showed some variability, the charge seemed not to vary continuously, but was an integral multiple of what he called e, the postulated charge on an electron. He concluded that the total charge on a droplet was an integer multiple of the number of discrete electrons in the droplet.

At about the same time, Felix Ehrenhaft of the University of Vienna in Austria was carrying out similar experiments. Ehrenhaft claimed that the variability in Millikan's published measurements supported Ehrenhaft's theory that there must be subelectrons, each carrying a charge that is a fraction of that carried by an electron. Millikan and Ehrenhaft were at odds scientifically, and the issue of subelectrons became a major issue of debate; it ended soon afterward among scientists of that era when people lost interest. The argument about subelectrons has been revisited in recent years, however, as it relates to the quark, the elementary subnuclear particle that was found recently. Related particles are postulated that have one-third and two-thirds of the charge of the electron. The subelectron debate may not be over after all.

In 1913, Millikan published new experimental results (Millikan, 1913) that he claimed were more accurate, and which purported to confirm his earlier conclusions. In italics, he stated: *"This is not a selected group of drops but represents all of the drops experimented upon during 60 consecutive days."*

That should have ended the debate and should have shown that through objective, scientific measurement the situation was completely clarified; but Millikan's perhaps excessively generous use of scientific subjectivity in the reporting of his results was discovered later, by Gerald Holton (1978), who scrutinized Millikan's laboratory notebooks. Holton went back to the original notebooks and submitted papers on which Millikan based his 1913 paper and found some major problems with Millikan's reporting of his data. Holton reported (1978, p. 52): "With idiosyncratic frankness and detail, Millikan shows in the section, 'The Results', that the measurements with the new technique were still difficult to make, that he relied heavily on personal judgment, and that it was really still his first major paper."

In short, Holton was reporting that Millikan had been excessively subjective in his methodology, that he had crossed over the line of what is acceptable scientific practice. Holton (1978) explained that Millikan had given each of his original sets of observations in the 1910 paper a personal quality-of-measurement rating. Millikan had used three stars to grade the quality of his measurements "best" (there were 2 such), two stars for "very good" (there were 7 such), one star for "good" (there were 10 such), and finally, "fair" received no stars (there were 13 with no stars). In his paper, Millikan recorded, in the (perhaps naive) interest of full disclosure, that he had discarded seven observations entirely. Then, in the mode of a modern data analyst who does not want to ignore any of the information available in an experiment, Millikan assigned weights to the various qualities of his observations. He then formed a weighted average of the assessments of e that he found in each set of observations of equal quality. His final estimate of e was the weighted average, $e = 4.85 \times 10^{-10}$ esu (electrostatic units). But the ordinary unweighted average (mean) would have been 4.70×10^{-10} esu.

Our interpretation of the weights he used is that each of them represents Millikan's prior probability (his personal degrees of belief; see Chapter 5) for each of the groups of observations that they were yielding the true values. Millikan also mentions in his paper (Millikan, 1909–1910, p. 220) a very subjective, excessively strong personal belief that transcends his objectivity in measurement—that he would discard some observations if they had not agreed with the results of his other observations: "*I would have discarded them had they not agreed with the results of the other observations*, and consequently I felt obliged to discard them as it was." Moreover, in his laboratory notebook for March 12, 1912, there are annotations that Millikan associated with some of his observations, terms such as "publish this," "beauty," and "error high, will not use." So Millikan had very strong beliefs about what he was expecting to find and ignored the data that disagreed with his prior beliefs.

It seems that although Millikan claimed to be reporting all of his measurements in his published paper, he had actually selected his data subjectively. His 1913 article presented 58 data points (made on consecutive days) selected from a total of 140. What happened to the remaining missing data points? This question was not fully answered. Millikan's assessed mean value of e in his 1913 paper was $e = 4.774 \pm 0.009 \times 10^{-10}$ esu (the currently accepted value for e is 4.77×10^{-10} esu).

When a scientist tries to learn from the earlier research of others by combining the results of their experiments in some reasonable, quantitative way, the procedure is called *meta-analysis* in modern terminology. It seems that Millikan performed an early version of meta-analysis before reporting his results in his *Philosophical Magazine* paper (1910). In the section of the paper giving "the most probable value of the elementary electrical charge" (see Holton, 1978, p. 51), Millikan presents his own new mean value, $e = 4.65 \times 10^{-10}$ esu. He also assigns equal weight to all the recent determinations of e by methods that seem least open to question:

1. The value obtained by Planck from radiation theory (4.69×10^{-10} esu), which Rutherford had mentioned with favor at Winnipeg
2. The value of Rutherford and Geiger (4.65×10^{-10})
3. E. Regener's value, obtained by a method similar to Rutherford's (4.79×10^{-10})
4. Begemen's recent and as yet unpublished value of 4.67×10^{-10}, obtained in Millikan's laboratory

The final mean for e, Millikan declares, is thus 4.69×10^{-10} esu. So Millikan suggested to the scientific community that it should accept as the true value of e a simple average of those of the values reported previously that in Millikan's view were tolerably acceptable, including, of course, the value he had found himself. Millikan's meta-analytic approach was of course fundamentally subjective in his choices of which (whose) values to include in the averaging.

Millikan's personal belief was that the electron was not divisible, so it carried an integral discrete charge. This belief was so strong that he blatantly admitted that he would have ignored data points that disagreed with his preconceived thesis by just not publishing them. As it has turned out, Millikan's final experimental value, confirmed by later experiments with x-ray diffraction by crystals, was about the value that is officially accepted today. Despite, or perhaps because of, his subjectivity, he seems to have gotten it right.

REFERENCES

Broad, William, and Nicholas Wade (1982). *Betrayers of the Truth*. New York: Simon & Schuster.

Encyclopaedia Britannica Micropedia (1975). Vol. VI, p. 898.

Holton, Gerald (1978). "Subelectrons, Presuppositions, and the Millikan–Ehrenhaft Dispute," pp. 25–83 in Gerald Holton, *The Scientific Imagination: Case Studies.* Cambridge: Cambridge University Press.

Millikan, Robert A. (1909–1910). "A New Modification of the Cloud Method of Measuring the Elementary Electrical Charge, and the Most Probable Value of That Charge", *Physical Review*, **29**:560–561 (abstract, 1909); the full paper was published in 1910 in *Philosophical Magazine*, pp. 209–228.

——(1913). "On the Elementary Electrical Charge and the Avogadro Constant," *Physical Review*, **2**:109–143.

3.5   CYRIL BURT (1883–1971)

On February 22, 1997, Ian Wilmut of the Roslin Institute, an animal research laboratory near Edinburgh, Scotland, made a startling announcement to an astonished world. By means of gene manipulation involving three distinct sheep (see Section 3.5.1 for a description), he and his collaborators had successfully created the first animal cloned from an adult mammal, a lamb named Dolly. [See Wilmut et al. (1997) for technical details, and Specter and Kolata (1997) for a good newspaper account.] The implications of this important research are enormous and multifaceted, including both the dream and the nightmare of the possible eventual cloning of human beings. There are very serious ethical, religious, political, and social considerations that must eventually be addressed in relation to this research, but let us set them aside for the moment. One clear implication of this line of research is the possibility of being able to study, for scientific research purposes, the various factors that relate to the nature–nature debate, by comparing characteristics of clones reared in the same environment with characteristics of those reared apart.

This nature–nurture debate was the subject of much of the research of the eminent British educational psychologist Sir Cyril Burt (1883–1971). Burt reasoned that the only sure way to distinguish inherited from acquired traits would be to study identical twins, pairs of people who originate from the same fertilized egg and thus originally would necessarily have identical genetic composition, and to study how their intelligence varies with environmental factors. Perhaps the cloning of human beings will never take place (although now such an eventuality appears to be closer to a possibility than at any time in the past). Nevertheless, the possibility arises of extending Burt's twin research to the study of characteristics in clones of animals, because they would have *almost* identical genetic makeup. (But would the genetic differences in clones cause even more trouble?; see Section 3.5.2).

Possibilities of such clone research might well be especially enticing because Burt's identical-twin research has posthumously been clouded in controversy over whether he fabricated his data in an overzealous attempt to provide "scientific evidence" that supported his strong personal subjective beliefs about heredity and intelligence. Because of the possibility of being able to study larger samples of animals with known genetic composition, humankind might finally be able to address questions about heritability in a less controversial way. [A recent discussion of such controversies may be found in Herrnstein and Murray (1994), a source that itself has invoked much controversy because some have interpreted its conclusions to be redolent of racism. Another new work on the subject was published by Wright in 1997; see also Devlin et al. (1997) and Gould (1981).] But now we focus on Cyril Burt and the subjectivity in his scientific research.

Cyril Burt was born in London, England, in 1883. He was educated at Oxford University, graduating in 1907. In his postgraduate years he came to be influenced by the newly developed school of *biometric methods* (statistical methods applied to problems in biology and especially in inheritance) of Sir Francis Galton, and of Karl Pearson at University College, London, and the work on understanding intelligence of Charles Spearman. Galton strongly believed in the effects of heredity on intelligence and called for study of how research on twins (2% of the population) could affect understanding of such hereditary effects (Galton, 1875). Galton had suggested that two-egg, fraternal (dizygotic) twins (75% of all twins) could be used to understand the variation of hereditary characteristics to be expected within a family, whereas single-egg, identical (monozygotic) twins (25% of all twins) could be used to demonstrate the effects of environment. Burt began research on intelligence; his first published paper (Burt, 1909) supported the hereditary determination of ability (Joynson, 1989, p. 21). Because his research in quantitative educational psychology involving juvenile delinquency, intelligence testing, and *factor analysis* (a statistical tool for analyzing multidimensional data, such as answers to questions on intelligence tests administered to many subjects) was considered to be pathbreaking, eventually his fame extended far beyond England. By the time he was knighted in 1946, he was a world-renowned scientific figure. As Hearnshaw wrote (1979, p. 227), "Burt's reputation survived his lifetime. Not long before his death (1971) the American Psychological Association bestowed on him the Edward Lee Thorndike award, a high honor which had never previously been accorded to a foreigner. . . . In the obituary in *The* (London) *Times Educational Supplement* (15 Oct., 1971), Burt had been described as 'Britain's most eminent educational psychologist', and in the Cattell Feschrift (Dreger, 1972), which came out a little later, he was called 'dean of the world's psychologists'. Nobody then anticipated the criticisms and onslaughts to which his work was shortly to be subjected."

Burt was no stranger to controversy. His work came under attack on several occasions during his lifetime, but he had always fought back and defended himself. But after his death, his work came under serious questioning yet again. The problem started with an article by Arthur Jensen (1969) about the Headstart Program in the United States. The Headstart Program was designed to help children who had been raised under culturally deprived conditions overcome educational handicaps by giving them a headstart through extra schooling. Jensen argued, using Burt's twin data, that the Headstart Program was doomed to failure from the beginning, because children's educational handicaps were properly attributable to heredity, not environment. Thus putting resources into Headstart was a waste of money, time, and energy.

Three years after Burt's death, the matter was taken up by Leon Kamin (1974). Kamin argued the other side of the issue—that it was not heredity but environment that was the culprit for educationally handicapped children. His reasoning was based on the notion that Burt's twin data on identical twins reared apart that Jensen had relied on for the support of his argument were highly questionable. Kamin's book was an attack on the entire school of psychologists who had argued throughout the twentieth century that it was heredity that was the cause of educational handicap. He was arguing that "the hereditarians had allowed their supposedly scientific conclusions to be biased by their political convictions" (Joynson, 1989, p. 29). Kamin, pointing out that the hereditarians' belief system included endorsement of ideas of eugenics, which encouraged selective breeding of humans to improve the stock, asserted that their subjective beliefs and judgments were not only intertwined with their experimental data, but perhaps the data had been altered to support the conclusions they sought.

Kamin's 1974 criticisms against Burt's twin research claimed:

1. That there was little in Burt's writing to explain precisely how the data were collected, which populations were tested, how test scores were adjusted, how parental IQs were obtained, the extent and duration of the separations of the twin pairs, or the precise ages of the twin pairs when their IQs were tested.
2. That there were conflicting and inconsistent statements about economic background, social class, and intelligence of the twin pairs, as well as statements about the numbers of children in institutions.
3. That there were careless errors in Burt's tables.
4. That whereas the numbers of twin pairs being studied in Burt's various reports, both monozygotic and dizygotic, changed from one report to another, many of the correlations remained the same, to three decimal places. For example, for monozygotic twins reared apart, Burt reported in 1955 that the group tests of intelligence yielded a correlation of 0.771 for 21 pairs, and also in 1966 for 53 pairs; for monozygotic twins reared together the correlation was 0.994 for 83 pairs in 1955, and the same for 95 pairs in 1966. These are exceedingly unlikely coincidences. There were 20 such coincidences in his table of 60 correlations, a result very difficult to accept as the legitimate report of findings.

The Kamin attacks were highly damaging to Burt's scientific reputation. But when Oliver Gillie claimed on the first page of the London *Times* (Gillie, 1976) that Burt, the father of British educational psychology, had "faked" his data, Burt was being publicly accused of outright fraud, and much of his earlier work then also came under suspicion of fraud. Some of Burt's earlier data, analysis, and conclusions on twins reared separately and together appeared in a now infamous paper (Burt, 1961) that argued for a heredity basis for IQ and social class. It was this paper in which Kamin had found unexplainable technical problems. That same Burt paper was very carefully reexamined in a 10-page article in *Science* by psychologist D. D. Dorfman (1978). Dorfman concluded that (p. 1186) "[t]he eminent Briton is shown, beyond reasonable doubt, to have fabricated data on IQ and social class."

The battle was joined. Burt had his apologists and his attackers, including many careful evaluators of his statistical methods. Some argued that while Dorfman had shown that Burt may have been sloppy or careless in various ways, there was no evidence that Burt had intentionally fabricated his data and that his results were fraudulent (Stigler, 1979). Others argued that Dorfman had no evidence for his claims against Burt, although there were indeed inconsistencies in the analyses. These inconsistencies could have been attributed to recording or computational errors, but might also have been attributable to Burt's trying to create the entries of tables from specified margins, as Dorfman had claimed (Rubin, 1978). In a joint letter to *Science*, the statisticians Rubin and Stigler (1979, p. 1205) agreed that "Burt's description of his work is very vague in many respects, and precisely because of this sloppiness it is impossible to determine, with the type of statistical investigation attempted by Dorfman, whether or not Burt fabricated data."

The case for and against Burt was reopened more recently by Joynson (1989) and Fletcher (1991). Jensen (1995, p. 2) suggests that on the basis of Joynson's and Fletcher's work, the case against Burt is "not proven." But after reconsidering all the evidence, reexaminations, arguments, and counterarguments, Mackintosh (1995, p. 145) concludes that "[i]t is difficult to resist the suspicion that Burt has something to hide, and that in some cases at least this ambiguity was deliberately designed to conceal serious inadequacy in the data—or worse."

It seems clear to us in the case of Sir Cyril Burt that once again, strong subjective beliefs on the part of a famous and successful scientist appear to have dominated his methodological approach to research. It appears that he reached the point where he may have stepped over the bounds of acceptable scientific practice to justify his own very strong personal beliefs that intelligence is largely attributable to heredity. He may or may not have gotten it right; only improved future studies will be able to decide the nature–nurture question.

3.5.1   Note 1—Cloning Dolly

Donor cells were taken from the udder of a pregnant Finn Dorcet ewe and treated to stop them from dividing. Meanwhile, an unfertilized egg cell was taken from a second sheep, a Blackface ewe, and its entire nucleus, along with its nuclear DNA, was sucked out. The denucleated egg cell was then placed next to a donor cell. An electrical charge placed on them caused them to fuse. A second electrical charge started the fused cell to grow by subdividing. Six days later the growing fused cell was implanted into the uterus of a third sheep, a Blackface ewe, where it grew to maturity. After gestation, a baby Finn Dorcet lamb named Dolly was born. An incredible aspect of the feat was that the already differentiated cell (one that had already gone through the process of deciding whether it would be muscle, hair, skin, legs, lungs, or whatever) from the udder of a mature sheep did not merely reproduce an udder, but rather, reproduced an entirely new, complete animal. Snakes that lose their tails can grow new ones; similarly, certain other animals can regenerate other organs. But in this case, cells from a mature mammal's udder produced an entire animal.

3.5.2   Note 2—Differences between Clones

In fact, the genetic makeup of clones would not be *precisely* the same, because of at least the following factors:

1. Eggs contain material other than the nucleus. "Eggs also contain structural and metabolic equipment, including a complement of extranuclear DNA specific to that individual; the second ewe did not contribute her nucleus, but she did contribute the rest of the contents of her egg" (Hubbard, 1997).

2. The genes don't really determine the structure of the brain. "The best the genes can do is indicate the rough layout of the wiring, the general shape of the brain" (Johnson, 1997, p. 1). After that, both the intrauterine environment and experience gained after birth work to shape the brain by making and breaking connections between neurons. "From the very beginning, what's in the genes is different from what's in the brain. And the gulf continues to widen as the brain matures" (Johnson, 1997, p. 1).

3. Even for ova that have split into two identical cells to form monozygotic (identical twin) fetuses, and therefore are gestated in the same uterus, the environment in which they develop is slightly different from the start.

Although identical twins are closer to one another than clones would be, because they share both intranuclear DNA, as well as the DNA included within the ovum that is outside the nucleus, they need not be identical in all of their body structure, let alone internal organs and brains. For example, identical twins need not have identical fingerprints, although the fingerprints of identical twins are likely to be closer to one another than randomly selected fingerprints would be. This is because some body cells undergo spontaneous mutations (at a very low rate), and because environmental influences affect such things as the timing of sequences of cell development, causing small differences in structural development, even in identical twins (Clegg, 1997). These mutational differences and environmental influences can produce different fingerprints in identical twins.

REFERENCES

Burt, C. L. (1909). "Experimental Tests of General Intelligence," *British Journal of Psychology*, **3**:94–177.

———(1961). "Intelligence and Social Mobility," *British Journal of Statistical Psychology*, **14**:3–24.

Clegg, Michael (1997). Professor of Genetics, Department of Botany and Plant Sciences, University of California, Riverside, personal communication, March.

Devlin, Bernie, Daniel P. Resnick, Stephen E. Fienberg, and Kathryn Roeder (Eds.) (1997). *Intelligence, Genes and Success: Scientists Respond to The Bell Curve*. New York: Springer-Verlag.

Dorfman, D. D. (1978). "The Cyril Burt Question: New Findings," *Science*, **201**:1177–1186.

———(1979). "Burt's Tables," Letters to the Editor, *Science*, **204**:246–254.

Dreger, R. M. (Ed.) (1972). *Multivariate Personality Research*, Chap. XI. Baton Rouge, LA: Claitor's Publishing Division.

Fletcher, R. (1991). *Science, Ideology, and the Media: The Cyril Burt Scandal*. New Brunswick, NJ: Transaction.

Galton, F. (1875). "The History of Twins As a Criterion of the Relative Powers of Nature and Nurture," *Fraser's Magazine*, **12**:566–576.

Gillie, Oliver (1976). "Crucial Data Was Faked by Eminent Psychologist," *Sunday Times* [London], October 24, p. 1.

Gould, Stephen Jay (1981). *The Mismeasure of Man*. New York: W.W. Norton.

Hearnshaw, L. S. (1979). *Cyril Burt: Psychologist*. London: Hodder & Stoughton.

Herrnstein, Richard J., and Charles Murray (1994). *The Bell Curve: Intelligence and Class Structure in American Life*. New York: Free Press.

Hubbard, Ruth (1997). "Irreplaceable Ewe," Editorial, *The Nation*, February 24.

Jensen, A. R. (1969). "How Much Can We Boost IQ and Scholastic Achievement?" *Harvard Educational Review*, **39**:1–123.

———(1995). "IQ and Science: The Mysterious Burt Affair," in N. J. Mackintosh (Ed.), *Cyril Burt: Fraud or Framed*. Oxford: Oxford University Press.

Johnson, George (1997). "Don't Worry. A Brain Still Can't Be Cloned," Week in Review, *New York Times*, March 2, p. 1.

Joynson, Robert B. (1989). *The Burt Affair*. New York: Routledge.

Kamin, L. J. (1974). *The Science and Politics of IQ*. New York: Wiley.

Mackintosh, N. J. (Ed.) (1995). *Cyril Burt: Fraud or Framed?* Oxford: Oxford University Press.

Rubin, Donald B. (1979). "Burt's Tables," Letters to the Editor, *Science*, **204**:245–246.

Rubin, Donald B., and Stephen M. Stigler (1979). "Dorfman's Data Analysis," Letters to the Editor, *Science*, **205**:1204–1206.

Specter, Michael, with Gina Kolata (1997). "After Decades and Many Missteps, Cloning Success," *New York Times*, March 3, p. 1.

Stigler, Stephen M. (1979). "Burt's Tables," Letters to the Editor, *Science*, **204**:242–245.

Wilmut, I., A. E. Schnieke, J. McWhir, A. J. Kind, and K. H. S. Campbell (1997). "Viable Offspring Derived from Fetal and Adult Mammalian Cells," Letter, *Nature*, **385**:810.

Wright, Lawrence (1997). *Twins and What They Tell Us About Who We Are.* New York: Wiley.

3.6 MARGARET MEAD (1902–1978)

Margaret Mead was a world-famous American anthropologist, best known for her studies of the nonliterate peoples of Oceania, especially with regard to various aspects of psychology and culture and the cultural conditioning of sexual behavior. She was also a popular celebrity, and in that world she was most notable for her forays into such far-ranging topics as women's rights, childrearing, sexual morality, nuclear proliferation, race relations, drug abuse, population control, environmental pollution, and world hunger.

During a 1925 field trip in Samoa relating to her doctoral thesis, she gathered material for the first of her 23 books, *Coming of Age in Samoa* (Mead, 1928; new ed. 1968), an example of her reliance on observation and a strong personal belief system, rather than on hard data. As we will see below, *Coming of Age* clearly reflected her strong belief in *cultural determinism*, a position that caused some later anthropologist critics to question both the accuracy of her observations and the soundness of her conclusions.

Coming of Age had long attained the status of an anthropological classic by the time of Mead's death in 1978. In this work, Mead claims to show that adolescence in Samoa is a stress-free stage of the life cycle, in stark contrast to the *sturm und drang* experienced by adolescents in the United States and in other Western cultures. Based on some five months of fieldwork, the book claims to show that the lack of stress Samoans feel in the transition from childhood to adulthood is traceable to two main factors. The first of these is the freedom that Samoan adolescents feel to indulge in sexual experimentation without feelings of guilt. The second factor is an upbringing that does not emphasize competitiveness and that involves residence in large extended families, so that intense emotional involvement between children and their parents is attenuated. All this was seen to give rise to an easygoing way of life, without intense emotions and involving a smooth transition from childhood to adulthood without a traumatic adolescence.

In 1983, Derek Freeman published *Margaret Mead and Samoa: The Making and Unmaking of an Anthropological Myth.* Freeman stressed that Mead was young and impressionable (she was 23) and poorly trained in ethnography (she had received very little explicit training in fieldwork, and none at all in the Samoan language until her arrival in the field). He pointed out that she was rushed in carrying out two major research projects in a very limited time and on a very limited budget. Freeman claimed that Mead reported seeing in Samoa what she had expected to see rather than what the society was actually like. That is, she had let her own beliefs dominate her methodological approach to a scientific investigation to such an extent that she, in part, either did not note or did not emphasize any data that did not support her preconceived hypotheses. Normally, social scientists develop understanding of the processes at work in the phenomena they are studying and combine that understanding with experimental data to help them to draw appropriate conclusions about a social process. In Mead's case, according to Freeman, she became excessively subjective in her scientific methodology and overstepped the bounds of scientific propriety.

During the 1920s the nature–nurture debate had been joined in the United States; at issue was whether human behavior was innate, biologically determined, and hence unchangeable by societal norms, cultures, and policies, or whether it was plastic, malleable, and hence influenced and influenceable by societies. (For more discussion of these issues, see, for example, the discussion of Cyril Burt in Section 3.5.) Because behavior is biologically determined, argued the eugenicists, societies could be improved by forced sterilization of "inferior" individuals and by restricting immigration of "inferior" ethnic and racial groups. These policies would make less sense if it could be shown that behavior is strongly influenced by culture.

Mead's mentor, Franz Boas, was one of the social scientific proponents of the nurture school; he charged Mead explicitly to find a culture that would support his position. (He originally suggested that Mead look for such a culture among American Indian tribes; she insisted upon going to Samoa.) If adolescence, which was thought to be "naturally" stressful in Western culture, could be shown to be free of stress under other cultural arrangements, this example would open the door to the consideration of whether other behaviors, heretofore seen as "natural," might not also be culturally determined. (Indeed, Mead's later work considered gender roles in a similar fashion. Starting out with the conception in Western culture of the feminine as passive, light-hearted, and nurturing and the masculine as aggressive, serious, and often warlike, Mead claimed to have found cultures in which all possible combinations of those gender stereotypes exist. In one, the Western roles were reversed, with men acting "feminine" and women "masculine"; in another, both sexes acted "feminine"; and in yet a third, both acted "masculine." She was following the logic encouraged by Boas and embodied in *Coming of Age in Samoa* of showing a counterexample of a social arrangement thought to be "natural," and hence undermining its claim to inevitability because of its naturalness.)

Freeman's book charged that Mead's limited time, limited resources, and limited familiarity with the Samoan language made it impossible for her to do the careful fieldwork that would make a judgment of the level of stress involved in adolescence scientifically valid. He himself spent about six years in Samoa between 1940 and 1980 (thus a generation later, and at a different part of the island than that at which Mead had spent her time), becoming fluent in the language, being adopted into a Samoan family, and being awarded the title of chief. Rensberger (1983, p. 33) summarizes Freeman's findings: "Where Mead saw an adolescence of ease and tranquillity, Freeman saw teenagers every bit as rebellious and troubled as those of the industrialized world. Where Mead saw a sexual freedom that banished everything from impotence and frigidity to rape . . . Freeman saw a puritanical attitude toward sex. . . ." Mead had insisted that rape was absent in Samoan culture. But not only does Freeman find rape in the Samoan culture he visited up to a half-century after Mead, he also finds frequent reports of rapes in the local papers dating from the time that Mead was in the field.

Further, on careful reading of Mead's own work, Freeman finds her noting exceptions, four adolescent girls whose behavior could be characterized as delinquent. Because the behavior of these girls was not consonant with the broad generalizations Mead was attempting to make about the tranquility of Samoan adolescence, she relegated their data to a separate chapter. Note here the down-weighting of data that do not fit neatly with the investigator's theory—similar to what we saw occurring with Kepler in Section 3.2 and Millikan in Section 3.4. Freeman also claims that as a joke, two of Mead's chief informants simply told her tall tales that they judged she wanted to hear. Indeed, Freeman secured an affidavit from one of these informants, Fáapuáa Fáamu, asserting that she had played a joke on the young anthropologist, not realizing that Mead would take it literally.

Freeman's book gave rise to a major controversy among anthropologists, with sessions at meetings and many publications, even leading to a compendium (Caton, 1990). An attempt to close the controversy was made by Orans (1996). As well as reading *Coming of Age in Samoa* carefully, Orans claims to have made a detailed examination of Mead's field notes and letters, and worked at reconciling the various sources. He first argues that Freeman stressed Mead's broad generalizations, ignoring her more guarded ones (such as the treatment of deviants in *Coming of Age in Samoa*). Orans sees the field notes as supporting a more restrictive view of adolescent girls' sexual milieu than does Mead's book. He also argues that Mead could not have been duped by Fáapuáa Fáamu's joking, as she was familiar with the Samoan custom of making such jokes. Further, she had already written field notes about the restrictions on ceremonial virgins (of which Fáapuáa Fáamu was one) before the "joking" took place. But the crux of Orans' argument is that Mead did not do any real science. Orans feels that she bases conclusions about behavior on the report of a single informant, and that there is no clear indication as to how she reached many of her conclusions about such attitudes as the lack of fear of divine retribution for premarital sex. For example, Orans (1996, p. 123) quotes Mead as saying "[there is] no frigidity, no impotence, except as the temporary result of severe illness, and the capacity for intercourse only once in a night is counted as senility." She argues that it is large families that produce less troubled adolescence, but she never compares behavior or attitudes between Samoan girls from larger families and those from smaller ones, although the data on family size were available to her from a census she herself carried out. And although she presents the pastor and his wife as influences against freedom of sexuality among the Samoan girls, she makes no comparison between those boarding with the pastor and those living with their own families. If she were really interested in contrasting Samoan adolescence with that in the United States, Orans argues, Mead would also have had to study American adolescents systematically or build on the work of someone who had already carried out such a study.

It is worth describing at length Orans' argument about the unscientific character of Mead's methods. Mead was aware of the standards of scientific rigor appropriate for the social sciences. Indeed, after about two months in the field, she wrote to her mentor, Boas (quoted by Orans, 1996, p. 125): "I ought to be able to marshall an array of facts from which another would be able to draw independent conclusions." But she despairs of doing so, because although she can tally cases, she knows she hasn't drawn a real probability sample: she has merely used her own judgment. Thus, a reader of her book is required to trust her judgment about what the data say, whether rendered "statistically" by presenting counts, or by the qualitative richness she greatly prefers.

Orans argues that Mead's sample need not have depended on her judgment; she could have drawn a random sample of female adolescents from the villages she studied. Orans summarizes tellingly (1996, p. 127):

> What Mead is claiming is that she is quite confident of her conclusions even though she has not done any formal sampling, and therefore the only reason to do such sampling would be to convince others. In addition, she claims a kind of insight based upon context and judgment that can never be justified by a presentation of observations. Her colleagues, then, must either except [sic] her judgment or not. Of course, even if one concedes that an observer may know more than she can demonstrate, one might still contend that a concession might more readily be made if one were presented with the observations that are available. Mead is asking to be excused from collecting and presenting a fair sample and supporting her conclusions with case histories. What she wants is permission to present data simply as "illustrative material" for the representativeness of which one will simply have to take her word.

Orans also quotes Boas' answer to Mead (1996, p. 128), in which he tells her not to attempt a statistical treatment, but writes: "A complete elimination of the subjective use of the investigator is of course quite impossible in a matter of this kind but undoubtedly you will try to overcome this so far as that is at all possible." It is not clear that Mead was successful in any attempt to overcome her apparently excessive subjectivity that resulted in inappropriate methodology. Rensberger (1983) points out that it is hardly unusual for cultural anthropologists to interpret the same culture differently because their preconceptions act as filters on their perceptions (in our terminology, their subjectivity influences both their data collection and interpretations).

That Mead used her judgment generously seems hardly in doubt. But she appears to have permitted her judgment not only to assist her in understanding the social processes she was observing, but to dominate her scientific methodology to such an extent that she appears to have manipulated her data so as to support her own hypotheses. Whether her excessive subjectivity permitted her to arrive at correct scientific conclusions is not clear. Derek Freeman insists that it led her to egregious error; Orans argues that her work was so totally unscientific as to be "not even wrong."

REFERENCES

Caton, Hiram (Ed.) (1990). *The Samoa Reader: Anthropologists Take Stock*. Lanham, MD: University Press of America.

Freeman, Derek (1983). *Margaret Mead and Samoa: The Making and Unmaking of an Anthropological Myth*. Cambridge, MA: Harvard University Press.

Mead, Margaret (1973 [1928]). *Coming of Age in Samoa*. New York: Morrow.

Orans, Martin (1996). *Not Even Wrong: Margaret Mead, Derek Freeman, and the Samoans*. Novato, CA: Chandler & Sharp.

Rensberger, Boyce (1983). "Margaret Mead: The Nature–Nurture Debate I," *Science83*: 28–37.

CHAPTER 4

Stories of Famous Scientists

4.1 INTRODUCTION

In this chapter we tell the stories of the 12 scientists who were selected for exploration in Jack Meadows' *The Famous Scientists* (see Chapter 2 for an explanation of how these scientists were selected) and therefore selected for more detailed examination in this book. The treatment is more extensive here than in Meadows; we discuss not only the lives of these people and their scientific contributions, we also discuss how they used their subjective views of the phenomena they studied to enhance the conclusions they drew. As in Chapter 3, the scientists we study in this chapter are all very famous and deservedly, highly respected. They include Nobel prize winners and others whose scientific contributions have been seminal in creating the structure of many fields of science. We have examined their lives and scientific contributions to study the informal ways they used their intuition, beliefs, and informed judgment to further their research.

We will see that in all cases, the informal use of judgment combined with very carefully obtained experimental data led these scientists to broadly based conclusions that shaped their fields for many years into the future. In a few instances, in a small portion of their scientific investigations, even these most-famous scientists may have stepped across the lines of acceptable methodological behavior in their confidence in their scientific judgment and in their quest for recognition by their peers and by the world at large. But the fabricating and fudging of data, the misrepresentation of scientific results, and the committing of other fraudulent practices is never acceptable, even by world-renowned scientists.

In Section 4.14 we provide some conjectures as to how and whether things might have been different had Bayesian statistical science been available to these famous scientists. We will see in Chapter 5 that by using Bayesian methods, the scientist is made very cognizant of where his or her judgments

49

and preconceived hypotheses enter the analysis. In this way the scientist can minimize the chances of biasing the experimental results while still being able to capitalize on informed knowledge and understanding of the underlying process.

4.2   ARISTOTLE (384–322 B.C.)

Aristotle was a philosopher, logician, naturalist, and scientist. His philosophical constructs formed the basis of Western thinking for the next 2000 years.

4.2.A   Brief Biographical Sketch

Aristotle was born in 384 B.C. in a region of what is now northern Greece, called Macedonia, then a city-state. He was the most prominent spokesman for the Realist school of philosophy in the history of Western philosophy (holding that *matter* is the only reality). Perhaps more than any other thinker, Aristotle has characterized the orientation and content of all that is termed *Western civilization.* His methodological approach to science was largely to observe, and then to generalize through enormous subjective speculation, without experiment. His subjective generalizations about science often turned out to be wrong (his conjectures sometimes lasted for 2000 years before being corrected), but he was also often correct. (By contrast, his philosophical assertions and proofs, especially those relating to the theory of logic, were not at all wrong, and have lasted to this day.)

Insight into why Aristotle developed as he did comes from understanding a bit about his background. Aristotle's father was a doctor who became court physician to the king of Macedonia, Amyntas III. In Aristotle's adult years, the throne of Amyntas III was held by Amyntas's son, Philip II; Philip's son was Alexander the Great. Aristotle was the son of a doctor, and as such, he was heir to a scientific tradition about 200 years old. Through case histories contained in the *Epidemiai* (Epidemics) of Hippocrates (about 440 B.C.), the father of medical science, Aristotle was introduced to Greek medicine and biology at an early age by his father, and this fact was greatly influential in determining many of his later interests. Medicine was traditional in certain families, being handed down from father to son by apprenticeship rather than by book learning. Thus, in all likelihood, Aristotle learned the fundamentals of that practical skill he was afterward to display in his biological researches.

While Aristotle was still a youth, his father died. The young man was then sent to Athens, where, at age 17, he became a pupil of the philosopher Plato, at his *Academy*, which may be called the first university. At the Academy, Aristotle was engaged in dialogue for 20 years, studying, writing, and eventually taking part in the teaching of philosophy, mathematics, ethics, politics, and aesthetics. Plato is said to have called him "the mind of the school." These years dominated Aristotle's intellectual development, under the formative influences of the philosophers Socrates and Plato; Plato had been a student of Socrates. Socrates and Plato, together with Aristotle himself, were destined to form a compact intellectual dynasty.

On Plato's death in 347 B.C., Aristotle left Athens. He went back north, this time to the coast of Asia Minor and the island of Lesbos. He again picked up his interest in natural science, and made notes especially on the structure and habits of the marine creatures he observed along the shore. At this time, Philip II was the king of Macedonia. Philip was a man of exceptional ability and driving will, who had for some years been at work transforming his realm from a poor, backward country of untapped natural resources and a population of rude huntsmen and mountaineers into a rich and powerful kingdom with an invincible army. The army had been organized to fight in solid blocks of heavily armed infantry, known as the *Macedonian phalanx*. Philip was seeking a distinguished Greek to take charge of the education of his only son, the fair-haired, high-spirited boy Alexander, then 13 years old. In 343 B.C. he offered the post to Aristotle, then 42, who accepted it.

In 340 B.C., Alexander, at 15, was made regent over the Macedonia Empire because his father, Philip II, was away in battle, in Greece. Because of Alexander's new responsibilities, he passed from Aristotle's surveillance. In 338 B.C., the armies of Athens and Thebes, which had been roused by Demosthenes (the great Athenian orator) to a futile defiance of Philip, were overwhelmed in the battle of Chaeronea. The rest of Greece succumbed gradually. Two years later, however, Philip was murdered, on the eve of launching his long-planned campaign against Persia, and Alexander, his son, reigned in his stead. Alexander soon showed that he had inherited a genius and ambition even grander than his father's. He promptly reduced all Greece to complete submission, then invaded and conquered, one after the other, all the empires of the Eastern Mediterranean—Assyrian, Egyptian, and Persian.

Not long after Alexander's assumption of power, Aristotle returned to Athens, then under Macedonian control. There he rented some buildings and opened a school known as the *Lyceum*. At the Lyceum, he collected one of the earliest reported libraries of manuscript books and scientific specimens of many sorts as material for research and lectures. Apparently, there was a botanical garden and a small menagerie of birds and other animals, perhaps because one or more of Aristotle's students accompanied Alexander on his expedition eastward and sent back accounts of the strange beasts and other things seen along the way. The chief difference in general character between Aristotle's new school and Plato's Academy was that the scientific interests of the Platonists centered on mathematics, whereas the main contributions of Aristotle's Lyceum lay, not surprisingly, in biology and history.

In 323 B.C., at the age of 32, Alexander died of a fever, and his generals divided his empire among them. Alexander's empire spelled the downfall of the small independent city-state that to Aristotle was the only possible form of civilized political community. Aristotle's Athenian activities and influence were brought to a hault by the early death of Alexander. The patriotic anti-Macedonian feeling that had festered just below the surface in Athens as long as the conqueror lived became overt after his death in an outbreak of hatred and bitterness against anyone with Macedonian ties or relationships. Aristotle, as Alexander's old tutor, who had come back to Athens in the wake of the Macedonian victory, was viewed with suspicion. A technical charge of impiety was lodged against him. Socrates had been treated the same way some 80 years earlier. But before the case against Aristotle could be brought to trial, he had fled to the north, refusing to let Athens "sin twice against philosophy." There, in Chalcis, a year later, he died, leaving a will in which he provided that several of his slaves should be freed. For over 800 years his pupils and their successors carried on the Lyceum. It was finally closed by order of a Christian emperor in Constantinople.

4.2.B   Aristotle's Scientific Contributions

We first discuss Aristotle's broad-ranging general contributions to civilization, and then focus on his involvement with what today we call *science*. To provide an inkling of Aristotle's enormous contributions to Western culture, we only need to define what is meant by *Aristotelianism*. Following the *Encyclopaedia Britannica* (1996), we see his influence in our language, with terms attributable to him ranging from *subject* and *predicate* in grammar and logic through terms applicable to science, such as *form, matter, energy, potential, quantity,* and *quality*. His philosophical methodology included the use of deductive reasoning proceeding from self-evident principles or discovered general truths. Aristotelian epistemology "includes a concentration on knowledge either accessible by natural means or accountable for by reason; an inductive, analytical empiricism, or stress on experience, in the study of nature—including the study of men, their behaviour and organizations—leading from the perception of contingent individual occurrences to the discovery of permanent, universal patterns; and the primacy of the universal, that which is expressed by common or general terms." Aristotle put forward the belief in four material elements and their basic qualities and the four causes: formal, material, efficient, and final.

Thus, we see that Aristotle has made profound and substantial contributions to civilization in his work in ethics, logic, philosophy, language, and metaphysics. Much of this work not only survives to the present day—it is current, practically in its original form. But what of his researches into science? There, the story is quite a bit different (see, e.g., Barnes, 1982, 1995; Grant, 1877; Solmsen, 1960; Veatch, 1974).

Given his early training in medicine, it is not surprising that among other things, Aristotle would become a naturalist and marine biologist and focus his observations on plants and animals. His biological writings constitute over 25% of the surviving corpus. In taxonomy, Aristotle was a careful and meticulous observer who classified animal species into various hierarchies of categories. He classified over 500 species of animals and dissected about 50 of them. For example, he observed that while the dolphin gave birth to live young and nursed them with its own milk after birth, fish did neither of these; so he classified dolphin with mammals instead of with fish. This classification was to be sustained 2000 years later.

In science, Aristotle seems to have conducted considerable questioning and observation, engaged in much conjecturing, made some enormous leaps of faith to generality probably far beyond justification by his observations, and often did substantial reasoning to erroneous conclusions, because he lacked appropriate observational data. With a few exceptions, the vast corpus of his scientific conclusions is today believed to be false. Part of the reason for what we see today as his failures in the experimental sciences stems from the unavailability of proper scientific instruments. He had no barometer, microscope, spectroscope, telescope, thermometer, or any of the countless other instruments we take for granted today. He could merely observe with naked eye and speculate.

Some of his ideas do survive, however. Although suggestions that the earth is not flat dated from Pythagoras (520 B.C.), Aristotle (and other Greeks earlier than Aristotle) cited the reasons that are still appealed to today. They recorded noticing that ships' masts and mountaintops disappeared last, and appeared first, on the horizon; they noticed that new stars appeared in the sky as people traveled farther south; and they noticed Earth's circular shadow move across the Moon during its eclipses. For these reasons, Aristotle and others before him correctly concluded that Earth must be spherical.

Aristotle also estimated the circumference of the earth to be 400,000 stades (an ancient Greek measure), or, equated to modern measures, 38,000 miles (Grant, 1877, pp. 138, 139). Today we believe that a better estimate is about 25,000 miles. But the error is not too bad for a 2300-year-old measurement. But Aristotle also argued that Earth is stationary in space. (It is, of course, in motion.) Moreover, he concluded that heavy objects move toward Earth with speeds proportional to their weights. (Galileo and Newton showed that all bodies move toward Earth at the same speed, in a vacuum; but this was about 2000 years later, so Aristotle's claim was the "scientific" result used for about 2000 years.)

In a biological context, Wyatt concluded (1924, p. 27):

> [Aristotle] alluded to the heart as the source as well as the reservoir of the blood. He distinguished between arteries and veins and said that they were complementary, each existing for the sake of the other. (*De Respiratione*, cap. ix)

> Further on he speaks of the heart as the seat and origin of the natural heat; and respiration, in his opinion, was instituted solely to cool this glowing organ and to moderate its heat and keep it within bounds. "The hotter the animal," he says, "the more vigorously must it breathe, in order the more effectually to subdue the heat; whence the larger development of the lungs in quadruped and birds than in amphibious animals." (Ibid., cap. xix)

> The movement or pulsation of the heart was due to the sudden expansion of the material supplied to it from the food digested in the stomach. When this came in contact with the innate heat of the heart, it expanded and caused that organ to dilate and give a beat. The supply of this material being continuous, the pulse accordingly was never still. (Ibid., cap. xx)

> The arteries pulsated at the same time as the heart did, and the blood flowed alternately from the vessels to the heart and from the heart to the vessels with a to-and-fro motion like the ebb and flow of the tides. The valves at the junction of the blood vessels with the chambers of the heart regulated these movements and alternately opened and shut; one movement of the heart opening one series of valves, the next movement closing these and opening others.

In 1627, William Harvey was to show Aristotle's understanding to be totally wrong, and that, in fact, the blood circulates (see the story of Harvey in Section 4.4). Aristotle's further misunderstanding of the anatomy of the heart is shown in his contention that the nerves derive therefrom. That misunderstanding has its basis in the long-standing belief that the heart is the seat of emotions, but such a misunderstanding is surprising in one who had carried out so many dissections. Grant (1887, p. 148) pointed out that in Aristotle's *On the Parts of Animals*, Aristotle "made a distinction, still valid in physiology, between tissues and organs." But Grant balances this view of Aristotle's getting it right by saying (p. 152) that "[i]t is hardly necessary to say that every opinion above mentioned is mistaken, and almost every statement of fact erroneous."

Aristotle's *History of Animals* remains an important work. According to Barnes (1982, p. 9), the title really means "Zoological Researches." He says:

> The *Researches* discusses in detail the parts of animals, both external and internal; the different stuffs—blood, bone, hair and the rest—of which animal bodies are constructed; the various modes of reproduction found among animals; their diet, habitat, and behavior.... It is easy to become starry-eyed over the Researches,

which are on any account, a work of genius and a monument of indefatigable industry. Not surprisingly, sober scholars have felt it incumbent upon them to point out the defects in the work.

First of all, it is said that Aristotle often makes errors of a crude and unscientific kind. Aristotle asserts more than once that during copulation the female fly inserts a tube or filament upwards into the male—and he says that "this is plain to anyone who tries to separate copulating flies." It is not. The assertion is wholly false.

Secondly, Aristotle is accused of failing to use "the experimental method." . . . There is no evidence that Aristotle ever attempted to establish correct experimental conditions or to make controlled observations; there is no evidence that he tried to repeat his observations, to check them, or to verify them.

Finally, Aristotle is criticized for ignoring the importance of measurement. Real science is essentially quantitative, but Aristotle's descriptions are mostly qualitative. He was no mathematician. He had no notion of applying mathematics to zoology. He did not weigh and measure his specimens.

Barnes mitigates these criticisms of Aristotle's work by pointing to Aristotle's lack of instruments, the fact that all researchers make errors of measurement, and that zoology is essentially a qualitative science. But the fact is that although Aristotle did show the way for science to develop, many of his own zoological beliefs and conclusions were wrong. Aristotle had no concept of chemistry. He decided subjectively that the world was composed of four elements: earth, water, air, and fire. Thales of Miletus 300 years earlier had claimed that there was only one such element, water. Today we are aware of over 100 elements, and water is not among them.

4.2.C Major Works

Aristotle's lectures at the Lyceum were eventually collected into almost a 150-volume set of encyclopedias encompassing the knowledge of his time, much of it Aristotle's own beliefs and observations. Only some 47 of these have survived through time, having been found in a pit, accidentally, in 80 B.C. by some soldiers who brought them to Rome, where they were copied.

Aristotle (1984). *The Complete Works of Aristotle*. Edited by J. Barnes. Princeton: Princeton University Press.

4.2.D Subjectivity in the Work of Aristotle

Aristotle's recorded scientific observations and experiences in biology were truly remarkable, in that they ushered in the entire approach to observational data that we use today and take for granted as if that were always the way of science; it was not, until Aristotle. But Aristotle's inferences about nature, and his conclusions about the physical world, were not always based on objective observation; they were often tendentious and speculative.

Aristotle's teacher, Plato, was a rationalist, believing that he could arrive at all knowledge about the universe by reasoning, and therefore, he was more than willing to rely on the deductive use of mathematics for his conclusions. He reasoned through *dialogues* (discussion and refutation) rather than *monologues*. Aristotle sometimes argued by reasoning and refutation, but by contrast, often tended to shun mathematics. He preferred instead to use the senses to observe, and to form opinions, based on empirical data and speculation (hypothesizing) about the meanings of the data, and then to express his views in peripatetic monologues, so-called because he walked around the Lyceum while explicating.

We can trace the influence of Aristotle's preconceptions on his observational work, not only from the questions he asked, but also from the answers he gave, as he represents the results of his research. His search for final causes (teleology) is an often-cited example, although it is not so much his general assumption of function and finality in biological organisms as some of his rather crude particular suggestions that are open to criticism. Moreover, he is clear that not everything in the animal serves a purpose and that it is not only the final cause that needs to be considered. But many slipshod or plainly mistaken observations (or what purport to be such) relate to cases where we can detect certain underlying value judgments at work.

The assumption of the superiority of right over left is one example. His repeated references to the differences between humans and other animals, and between males and females, are two other areas where errors and hasty nonobjective generalizations are especially frequent. Humans are not only marked out from the other animals by being erect—by having their parts, as Aristotle puts it, in their natural positions—and by possessing the largest brain for their size, hands, and a tongue adapted for speech. Aristotle also claims, more doubtfully, that human blood is the finest and purest, that human flesh is softest, and that the male human emits more seed, and the female more menses, in proportion to their size. While Aristotle's general distinction between male and female animals relates to a capacity or incapacity to fabricate the blood, he subjectively records largely or totally imaginary differences in the sutures of the skull, in the number of teeth, in the size of the brain, and in the temperature of men and women.

His view that, in general, males are better equipped with offensive and defensive weapons than females is one factor that leads him to the conclusion that the worker bees are male (today we know that they are female). To be sure, he sometimes notes exceptions to his general rules. For example, he remarks that although males are usually bigger and stronger than females in the nonoviparous animals, the reverse is true in most oviparous quadrupeds, fish, and insects. Although he goes along with the common belief that male embryos usually move first on the right-hand side of the womb, females on the left, he remarks that this is not an exact statement because there are many exceptions. There are certainly many inaccuracies in his reported observations besides those that occur when an a priori assumption is at work.

We can document the influence of Aristotle's overarching theories on what he reports that he has seen. But there are other occasions when the theories themselves appear to depend on, and may in some cases even be derived from, one or more observations (however accurate or inaccurate) which accordingly take on a particular significance for his argument. Undoubtedly, the most striking example of this is his often-repeated statement that the heart is the first part of the embryo to develop. He says that the primacy of the heart is clear not only according to argument, but also according to perception. In reporting his investigation of the growth of hen's eggs (a completely *uncontrolled* experiment) he notes that (*Historia Animalium*, vi, 561a6ff., 11–12) after about three days, the heart first appears as a blood spot that "palpitates and moves, as though endowed with life." The circumstantial detail of this and other accounts show that they are based on firsthand inspection, although the conclusion Aristotle arrived at is not entirely correct. As Ogle (1897) put it, "the heart is not actually the first structure that appears in the embryo, but it is the first part to enter actively into its functions." Lloyd (1987, p. 58) traced the results of Aristotle's observation: "This provides the crucial empirical support for his doctrine that it is the heart—rather than say the brain—that is the principle of life, the seat not just of the nutritive soul, but also of the faculty of locomotion and of the common sensorium."

Aristotle apparently frequently drew his conclusions from the opinions of others. He would ask other people, fishermen, hunters, and other passersby what they believed about something and would use those proxy views as his observational data, rather than his own, firsthand observations. Not surprisingly, conclusions drawn from these proxy observations were often wrong.

In both his physical and biological treatises, Aristotle often created general theories largely by projection from very small samples of data, if any. Eventually, with the aid of scientific measuring instruments developed almost 2000 years later, the fields of astronomy, physics, and anatomy as they were developed by Copernicus, Galileo, Newton, and Harvey proved Aristotle mistaken on so many points that his credibility as a naturalist began to come under question as well. In his research into the physical world he goes wrong more often than not and draws conclusions that are mostly subjectively based. But in his studies of biology, where he was better prepared for the scholarly approach, he works to somewhat better success. In his work on logic, his scientific contributions are as fresh today as ever. But the inescapable conclusion is that Aristotle the scientist depended, to an enormous extent, upon subjective judgment, intuition, observations of others, and creativity, mixed with some careful personal observations and some often hazardous leaps of faith.

REFERENCES

Aristotle, Edited (1943). Introduction by Louise Ropes Loomis. New York: Gramercy Books.

Asimov, Isaac (1989). *Asimov's Chronology of Science and Discovery.* New York: Harper & Row.

Barnes, Johnathan (1982). *Aristotle.* Oxford: Oxford University Press.

———(Ed.) (1995). *The Cambridge Companion to Aristotle.* Cambridge: Cambridge University Press.

Encyclopaedia Britannica (1975). Vol. 1, pp. 1162–1171.

———(1996). "Aristotle and Aristotelianism: Assessment and Nature of Aristotelianism," compact disk version.

Grant, Sir Alexander (1877). *Aristotle.* Edinburgh: William Blackwood and Sons.

Lloyd, G. E. R. (1987). "Empirical Research in Aristotle's Biology," pp. 59ff. in Allan Gotthelf and James Lennox (Eds.), *Philosophical Issues in Aristotle's Biology.* Cambridge: Cambridge University Press.

Ogle, W. (Trans.) (1897). *Aristotle on Youth and Old Age, Life and Death and Respiration.* London: Longmans.

Russell, B. (1959). *The Wisdom of the West.* New York: Doubleday.

Solmsen, Friedrich (1960). *Aristotle's System of the Physical World: A Comparison with His Predecessors.* Ithaca, NY: Cornell University Press.

Veatch, Henry B. (1974). *Aristotle: A Contemporary Appreciation.* Bloomington, IN: Indiana University Press.

Wyatt, R. B. Hervey (1924). *William Harvey.* London: Beonard Parsons; Boston: Small, Maynard.

4.3 GALILEO GALILEI (1564–1642)

Galileo Galilei was an Italian mathematician, astronomer, and physicist who advocated that experiments be included in scientific methodology; he qualitatively enunciated the physical laws of motion; and he exploited the use of the recently discovered telescope by discovering the moons of Jupiter and the rings of Saturn.

4.3.A Brief Biographical Sketch

Galileo was born at Pisa on February 15, 1564, the son of Vincenzo Galilei, a musician, and Guiulia Ammanati. His first name was given according to the local custom of doubling the family name for the firstborn. The family returned to Vincenzo's native Florence in 1574, and in the following year Galileo began his formal education at the monastery of Vallombrosa nearby. After about four years, the young man decided to become a monk, but his father intervened and spirited him away from the monastery. In 1581 he entered the University of Pisa to study medicine. He showed scant interest in the required courses but a great deal of interest in the peripheral course of mathematics. He began to study mathematics and science with Ostilio Ricci, a teacher in the Tuscan court. Observing the regularity of a pendulum swinging in church, Galileo devised a device to measure the human pulse, thus making a temporary good impression on his professors of medicine, who had been despairing at his lack of attendance at lectures and his argumentative ways when he did attend. But in 1585, the professors' dislike of Galileo persuaded them to deny him a scholarship and thus he had to withdraw from the university because of lack of funds before he had received a degree. Returning to Florence, he helped to support the family by tutoring private students in mathematics and even by returning to the monastery at Vallombrosa to teach the novices about mathematical perspective.

In 1586, at the age of 22, Galileo produced his first published scientific work, an essay describing the hydrostatic balance, the invention of which made his name known throughout Italy. He was eager to obtain a chair of mathematics at a university. In 1587 he applied for such a post in Bologna but was not selected; in 1589, however, he won a three-year appointment to the post of mathematics lecturer at the University of Pisa. He was very poorly paid, both because he was an academic beginner and because mathematics was considered an inferior discipline that did not command high salaries or prestige. Galileo developed a contempt for other members of the faculty who were Aristotelians (see Aristotle, Section 4.2) and who took the received wisdom of the master as unquestionable truth. His research into the theory of motion—in which he surely carried out both actual experiments and thought experiments, and may have experimented by dropping balls off the tower of Pisa—disproved the Aristotelian contention that bodies of different weights fall at different speeds. His contract was not renewed.

He then mounted a campaign for appointment to the chair of mathematics at Padua, a campaign that involved traveling to the seat of the Serene Republic of Venice to meet with a board of examiners and to lobby for support of the appointment in the Venetian Senate. His appointment went into effect in the fall of 1592; he was to remain at Padua for 18 years and perform the bulk of his most outstanding work there. At Padua he continued his research on motion and proved theoretically (about 1604) that falling bodies obey what came to be known as the law of uniformly accelerated motion (in such motion a body speeds up or slows down uniformly with time). He also gave the law of parabolic fall; for example, a ball thrown in a vacuum follows a parabolic path.

In 1594 and 1595 Galileo lectured on Ptolemy (who, in about 140 A.D., had advanced the geocentric theory that the Sun and planets revolve around Earth; see Kepler, Section 3.2), thus establishing himself as an astronomer. But Galileo had became convinced early in life of the truth of the heliocentric Copernican theory (i.e., that the planets revolve about the Sun)—he expressed support only privately, however. In 1597 he gave an exposition of the theory to an old friend Jacopo Mazzoni, and in the same year he responded to a book sent to him by Johannes Kepler that he had been a Copernican for many years. He felt, however, that public support would inspire ridicule. During these years Galileo had increasing family responsibilities. His father had died, leaving to his eldest son such tasks as assisting in the support of a younger brother hoping to make a career as a musician and paying the dowries for his sisters. In addition, Galileo began a liaison with Marina Gamba which was not to result in marriage but which lasted 10 years and produced three children.

In 1604 a new star appeared in the sky; Galileo gave a series of three public lectures about it in Padua, his first public appearance as an astronomer. Any new star, let alone one that moved, constituted a challenge to the established order of things, which held that the heavens were perfect and unchanging. Galileo accepted the evidence of his senses and spoke of the movement of the star—and in a published discourse on the phenomenon even hinted that Earth might also be moving.

While in Venice in the spring of 1609, Galileo learned of the recent invention of the telescope, and after returning to Padua he built one of threefold magnifying power, quickly improving it to a power of 32. This began a period of frantic activity. He produced numerous telescopes for his own use and to offer to influential people as gifts. Within the next two years he made and announced a series of astronomical discoveries. He found that the surface of the Moon was irregular rather than smooth, as had been supposed, and likened it to the surface of Earth, with seas and mountains. He observed that the Milky Way was composed of a collection of stars. He discovered four of the satellites of Jupiter and named them the Medican Stars, hoping to gain favor with his former pupil and future employer, Cosimo di Medici II, Grand Duke of Tuscany, and his illustrious family. He observed sunspots, the phases of Venus, and the rings of Saturn. His first astronomical observations were published in 1610 in *Sidereus Nuncius* (The Starry Messenger).

These important discoveries resulted in an appointment by the Grand Duke of Tuscany as "first philosopher and mathematician" at the court in Florence, a post that was free of teaching responsibilities and which Galileo accepted in 1610. In 1611, fearing that his enemies were plotting against him, Galileo traveled to Rome to win papal support for his work. He demonstrated his telescope to the most eminent personages at the pontifical court, including the pope himself. While he was in Rome, he was also elected to membership in the Academy of the Lynx, reputed to be the first scientific academy, founded and headed by Prince Frederico Cesi. The Lincean Academy in the future was to provide him with information on how his work was being received, support for his cause, and a publication outlet for his best-known work.

Returning to Florence, Galileo prepared and delivered a debate (and thereafter a treatise) on floating bodies, perhaps his first foray into experimental physics and the basis for his future work on mechanics. He learned in late March 1612 that a stargazer using the pseudonym of "Apelles" had observed sunspots, claimed priority for their discovery, and attributed them to the movement of stars across the face of the Sun; Galileo had been observing sunspots for 18 months. To assert his priority, he wrote three letters on the sunspots and had them printed at Rome in 1613. In these letters he took a more definite position on the Copernican theory, using the movement of the spots across the Sun as evidence for that theory. Because he wrote these letters in Italian rather than the usual Latin of academic discourse, they became well known and broadly influential. Angered because they had vested interests in the Ptolemiac theory, the professors at the university denounced him to the Inquisition, citing the contradictions between Copernican theory and Scripture. What ensued is succinctly described in the *Encyclopaedia Britannica* (1975, p. 852):

> Gravely alarmed, Galileo . . . wrote letters meant for the Grand Duke and for the Roman authorities . . . in which he pointed out the danger, reminding the church of its standing practice of interpreting Scripture allegorically whenever it came into conflict with scientific truth, quoting patristic authorities and warning that it would be "a terrible detriment for the souls if people found themselves convinced by proof of something that it was made then a sin to believe." He even went to Rome in person to beg the authorities to leave the way open for a change. A number of ecclesiastical experts were on his side. Unfortunately, Cardinal Robert Bellarmine, the chief theologian of the church, was unable to appreciate the importance of the new theories and clung to the time-honoured belief that mathematical hypotheses have nothing to do with physical reality. He only saw the danger of a scandal, which might undermine Catholicity in its fight with Protestantism. He accordingly decided that the best thing would be to check the whole issue by having Copernicanism declared "false and erroneous" and the book of Copernicus suspended by the congregation of the Index. The decree came out on March 5, 1616. On the previous February 26, however, as an act of personal consideration, Cardinal Bellarmine had granted an audience to Galileo and informed him of the forthcoming decree, warning him that he must henceforth neither "hold nor defend" the doctrine, although it could still be discussed as a mere "mathematical supposition."

It should be understood that Galileo's stance, if substantiated, would pose a real problem for the church; it might require a whole new Biblical science that would have to be put into place essentially overnight. Far more efficient to prohibit publication of such troublesome work.

Galileo returned home to Bellosguardo near Florence, close to where his daughters had now entered the convent of the Poor Clares, and continued his studies. He computed tables of the motions and eclipses of the moons of Jupiter. In 1618–1619 three comets appeared and caused a great stir. Galileo was too ill to make careful firsthand observations but was dismayed to read the publication of a Jesuit, Horatio Grassi, that declared the comets to be off-spring of Mercury and located between the Moon and the Sun, thus tending to discredit the Copernican system. A disciple of Galileo's, Mario Guiducci, wrote a reasoned rebuttal, but Galileo, still smarting about his priority on the sunspot issue, added sharp jabs at "Apelles" and at the entire Jesuit order. Grassi replied in kind, and in 1623 Galileo's reply, "The Assayer," appeared. It was a brilliant polemic on physical reality and an exposition of the new scientific method. In it he distinguished between the primary (i.e., measurable) properties of matter and the others (e.g., odor) and wrote his famous pronouncement that the "Book of Nature is . . . written in mathematical characters." Also in 1623, Maffeo Barberini, who had been a longtime friend of Galileo, became Pope Urban VIII, and Galileo dedicated the book to him. Pope Urban received the dedication enthusiastically.

But when Galileo went to Rome in 1624 to ask the new pope for permission to write about the Copernican theory, the permission was given only conditionally. He could write about both systems, Ptolemaic and Copernican, treating the latter hypothetically and coming to the appropriate conclusion— that humans cannot presume to know how the world is really made because God could have brought about the same effects in ways that we cannot imagine. Humans must not restrict God's omnipotence. Galileo used this permission to write his great book *Dialogue Concerning the Two Chief World Systems—Ptolemaic and Copernican*. In it, two savants and a third character, Simplicio, discuss cosmology; the weight of argument supports the Copernican position. Nevertheless, with a preface and conclusion written by the papal censor, the book received a full imprimatur.

Before long the pope learned the true position of the book as a compelling and unabashed plea for the Copernican system—indeed, the censor's conclusion at the end seemed not to fit the argument at all. Others whispered to the pope that the character of Simplicio was modeled on him; the pope was furious and ordered a prosecution. But because Galileo had received an imprimatur, there seemed no ground for prosecution. At this point a document was found in Galileo's Inquisition file that said that during his audience with Bellarmine on February 26, 1616, Galileo had been specifically enjoined from "teaching or discussing Copernicanism in any way." (Modern historians believe the weight of evidence suggests that this document had been manufactured after the fact to meet the need for a basis for prosecution.) Thus the imprimatur had been received under false pretenses and Galileo could be prosecuted for disobedience and "vehement suspicion of heresy."

After delays caused by his illnesses, Galileo went to Rome in February 1633 to stand trial. In deference to his age and celebrated works, he was not jailed but stayed first in the palace of the Tuscan ambassador and later in the Vatican itself. Under interrogation he denied any memory of the 1616 injunction. Nevertheless, Galileo was found guilty of having "held and taught" the Copernican doctrine and was ordered to recant and then remain imprisoned. After spending some months as the guest/prisoner of the Archbishop of Sienna, he was held under house arrest at his home at Arcetri, near Florence, from December 1633 until his death in 1642.

Even under house arrest and despite the failing of his never-robust health, Galileo continued to produce masterpieces. In 1634 he completed *Discourses on Two New Sciences*, in which he recapitulated the results of his early experiments and his mature meditations on the principles of mechanics. His last telescopic discovery—that of the Moon's diurnal and monthly librations (wobbling from side to side)—was made in 1637, only a few months before he became blind. Even after his blindness was complete, he worked out verbally a sketch of how a pendulum could be used to regulate a clock. He died of a

slow fever at Arcetri on January 8, 1642. In 1992 the Church finally exoner-
ated Galileo.

4.3.B Galileo's Scientific Contributions

In an objective sense, Galileo's achievements can be listed according to his
writings and the principles he enunciated. Thus, he is credited with developing "the empirical, or experimental, scientific method." His method combined
experiment with calculation, using experiments to check theoretical deduc-
tions. There is some controversy about Galileo's use of the experimental
method (see, e.g., Shapere, 1974); this issue is discussed below. In any case, his
discoveries described below resulted from using some combination of received
wisdom; reasoning using mathematics; and the telescope, the pendulum, and
the inclined plane:

1. He informally stated (in nonmathematical form) what came to be known
    (in their mathematical forms) as Newton's first two laws of motion: First,
    the principle of inertia—a body continues in a state of rest or uniform
    rectilinear motion forever unless acted upon by an external force; and
    second, that a body speeds up or slows down uniformly with time.
    Note that Galileo concluded that the rate of fall is independent of the
    constitution of the falling body; all bodies fall at the same rate, at least
    in a vacuum.
2. The fact that the path of a projectile in a vacuum is a parabola.
3. Numerous astronomical discoveries (including some of the moons of
    Jupiter, the rings of Saturn, the irregularity of the surface of the Moon,
    and the existence and behavior of sunspots) which together supported
    the Sun-centered Copernican system of the universe against the
    Ptolemaic Earth-centered system.

He also introduced the application of mathematics to the study and solution of scientific problems, stating that "the Book of Nature is written in
mathematical characters." As a consequence of these contributions, Galileo
often is referred to as the founder of modern mechanics and experimental
physics. He was the first to grasp the idea of "force" as a mechanical agent.
The concept of force was later elaborated mathematically by Newton, who
defined it in his second law of motion (see Newton, Section 4.5). Although
Galileo did not formulate the laws of motion per se, his writings on dynamics
clearly show that he understood and used the laws; in this sense he paved the
way for Newton in the seventeenth century.

Galileo's linking of mathematics and physics made it possible for him to
link celestial and terrestrial phenomena, and it broke down for him the traditional division of the world above the Moon and that below it. Galileo also
made some valid predictions of modern discoveries. He predicted that a small
annual parallax would eventually be found for some of the fixed stars that

is, a given star would be seen to move somewhat over the course of a year. He also predicted that extra-Saturnian planets would at some future time be ascertained to exist and that light travels with a measurable, although extremely great velocity.

4.3.C   Major Works

The first complete edition of Galileo's works was that of Eugenio Alberi, 16 vols. (1842–1856). It is now superseded by the final *Edizione nazionale delle opere di Galileo Galilei*, 20 vols., edited by Antonio Pavaro (1890–1909, reprinted 1929–1939). Florence: Barbera. It contains every obtainable document and scrap of correspondence. The first English translation of Galileo's writings was by Thomas Salusbury, *Mathematical Collections and Translations*, 5 vols. (1661–1962). London. Other English translations include:

Galilei, Galileo (1590). *De Motu* (unpublished manuscript). Translated by I. E. Drabkin
  as "On Motion," pp. 13–114 in *Galileo on Motion and on Mechanics* (1960).
  Madison, WI: University of Wisconsin Press.
———(1610). *Sidereus Nuncius.* Translated by Edward Stafford Carlos as *The Sidereal
  Messenger of Galileo Galilei and a Part of the Preface to Kepler's Dioptrics* (1880).
  London.
———(1612). *Discourse on Bodies in Water*, with introduction and notes by Stillman
  Drake (1960). Urbana, IL: University of Illinois Press.
———(1613). *Letters on Sunspots*, abridged text translated by Stillman Drake,
  pp. 87–144 in *Discoveries and Opinions of Galileo* (1957). Garden City, NY:
  Doubleday.
———(1623). *The Assayer*, translated by Stilllman Drake, pp. 151–336 in *The Controversy on the Comets of 1618* (1960). Philadelphia: University of Pennsylvania
  Press.
———(1632). *Dialogue Concerning the Two Chief World Systems—Ptolemaic and
  Copernican*, translated by Stillman Drake, 2nd ed. (1967). Berkeley, CA: University of California Press; in the Salusbury translation, Giorgio de Santillana (Ed.),
  Dialogue on the Great World Systems (1953). Chicago: University of Chicago Press.
———(1634). *On Mechanics*, translated by Thomas Salusbury as "Galileus, His
  Mechanics" pp. 271–302 of Vol. 2 of *Mathematical Collections and Translations*
  (1662). London. Translated by Stillman Drake, pp. 147–182 of *Galileo on Motion
  and on Mechanics* (1960). Madison, WI: University of Wisconsin Press.
———(1638). *Dialogues Concerning Two New Sciences*, translated by Henry Crewe
  and Antonio de Salvio (1914, reprinted 1954). New York: Macmillan; 2nd ed., New
  York: Dover. Also translated by Stillman Drake as *Two New Sciences* (1974).
  Madison, WI: University of Wisconsin Press. 2nd ed. (1989). Toronto: University of
  Toronto Press.

4.3.D   Subjectivity in the Work of Galileo

Galileo is popularly credited with being the first to develop and use the *scientific method*, and he is given as the exemplar of that method. (Aristotle proposed that we should first take observations. Francis Bacon borrowed from Aristotle, suggesting that this was really the only way science could advance. Galileo took the approach further by suggesting that we can carry out experiments to validate the theory that follows observations and that we use mathematics to predict experimental results.) Shapere (1974) credits the nineteenth-century philosopher-scientist Ernst Mach as a primary proponent of this view, quoting him (p. 4) as writing that Galileo "did not supply us with a *theory* of the falling of bodies, but investigated and established, wholly without preformed opinions, the *actual facts* of falling" and that Galileo proceeded by "gradually *adapting* . . . his thoughts to the facts."

Shapere also guides us through some of the voluminous literature that has been generated by historians and philosophers of science, exploring Galileo's role in the scientific revolution and attempting to understand his methodology. In particular, what is at issue is the role of theory (especially mathematics) and of experiment in Galileo's science. Shapere (1974, p. 129) quotes Alexander Koyre (1968) as taking the extreme opposite view from that held by Mach: "It is thought, pure unadulterated thought, and not experience or sense-perception, as until then [i.e., in the Aristotelian tradition], that gives the basis for the 'new science' of Galileo Galilei. . . . [According to Galileo] good physics is made *a priori*. Theory precedes fact."

Part of this confusion, perhaps, can be explained by the fact, cited by Miller (1996), that like Einstein many centuries later, Galileo performed "thought experiments" in addition to the actual physical experiments he carried out. An example is the work on the speed of falling bodies. The issue is that because of air resistance, in the real world very light bodies do indeed fall more slowly than do heavy ones; it is only in a vacuum that all bodies fall at the same speed. But there were no facilities to create laboratory vacuums in Galileo's time, so he had to imagine what would happen in a vacuum. Miller (1996, p. 8) describes one of Galileo's thought experiments on this subject: Consider a heavy rock that falls with 8 units of speed and a lighter one that falls with 4 units of speed. Now consider tying the two rocks together. How fast will they fall? Aristotelian principles dictate that the speed be somewhere between 4 and 8 units, as the lighter rock will impede the speed of the heavier. But Galileo asks how the rocks can know they are tied together, and whether it isn't possible that the combination will fall at a rate greater than 8 units.

Galileo's own writings are equivocal in resolving the question of whether he used experiments, and if so, whether they were exploratory or confirmatory. This ambiguity is in part because so many of his works were cast in the form of dialogues as an expository technique designed to appeal to a mass public and to make it possible for him with relative impunity to make statements that might be interpreted as challenging to church doctrine. Hence his writings can be mined for quotations supporting both sides of the controversy. For example, Shapere (1974, p. 126) presents two quotes from *Dialogue Concerning the Two Chief World Systems*. The first (from p. 223 of the 1962 Drake translation) says: "[I]n order not to proceed arbitrarily or at random, but with a rigorous method, let us first seek to make sure by experiments repeated many times how much time is taken by a ball of iron, say, to fall to earth from a height of one hundred yards." The second (from p. 328 of the Drake translation) has become famous:

> Nor can I ever sufficiently admire the outstanding acumen of those who have taken hold of this [Pythagorean, i.e., Copernican] opinion and accepted it as true; they have through sheer force of intellect done such violence to their own senses as to prefer what reason told them over that which sensible experience plainly showed them to the contrary. For the arguments against the whirling of the earth which we have already examined are very plausible, as we have seen; and the fact that the Ptolemaics and Aristotelians and all their disciples took them to be conclusive is indeed a strong argument of their effectiveness. But the experiences which overtly contradict the annual movement are indeed so much greater in their apparent force that, I repeat, there is no limit to my astonishment when I reflect that Aristarchus and Copernicus were able to make reason so conquer sense that, in defiance of the latter, the former became mistress of their belief.

Here is surely an argument that suggests that if the data contradict a theory, the data, not the theory, should be questioned.

Shapere (1974) also discusses the distinction between the experimental method used in an inductive mode to generate hypotheses and the method used in a deductive mode to test hypotheses otherwise generated or even to illustrate propositions already believed to be true. That author carries out a careful examination of the development of Galileo's conclusions that the distance traveled by a naturally falling body is proportional to the square of the time of fall. Galileo's first statement on this issue was in *On Motion*, completed around 1590, where he states that "a falling body will accelerate until the impetus that sent it aloft has diminished to zero; thereafter it will fall with constant speed." (Of course, we know today that objects do not fall at constant speed.) But Galileo changed his mind. Shapere (1974, p. 80) quotes Drake's translation of Galileo's *Fragment* dated 1604 as stating: "I suppose (and perhaps I shall be able to demonstrate this) that the naturally falling body goes continually increasing its velocity according as the distance increases from the point from which it parted."

His reasoning in that work, however, was based on deriving the fact that the distance fallen is proportional to the square of the elapsed time from the false proposition that velocity is proportional to distance traveled. This is evidence, suggests Shapere, that "Galileo was convinced of the truth of the time-squared relationship on grounds *independent* of the [velocity proportional to distance] relationship and the deduction of the former from the latter. Indeed, his search for a proof clearly came *after* his conviction of the truth of the conclusion, of which he was more firmly convinced than he was of the premise from which he attempted to derive it." Shapere goes on to consider explanations for Galileo's belief in the time-squared relationship as received wisdom, as a generalization of his experimental work with pendulums and inclined planes, or as arising from a process of pure reasoning.

In addition, as we have already seen in the case of a vacuum, there are questions of whether it was possible, in terms of laboratory equipment and techniques, for Galileo to have carried out the experiments he describes and that are sometimes credited to him. Many historians and philosophers of science have weighed in with opinions on this controversy, which is essentially about the roles of rationalism (with its roots in Plato's thought) and empiricism in the work of Galileo. In particular, Albert Einstein, in a preface to Drake's translation of *Dialogue Concerning the Two Chief World Systems* (written in 1952, published in 1967) argues for both the presence of subjectivity in Galileo's work and for its usefulness: "There is no empirical method without speculative concepts and systems . . . no speculative thinking whose concepts do not reveal . . . the empirical material from which they stem. . . . Moreover, the experimental methods at Galileo's disposal were so imperfect that only the boldest speculation could possibly bridge the gaps between the empirical data."

Einstein concludes that Galileo was aiming more at comprehension than at factual knowledge, and that comprehension depends on an already accepted logical system. That an already accepted logical system could affect Galileo's acceptance of others' ideas is illustrated by his neglect of Kepler's laws of planetary motion, which were discovered during his lifetime. But even while his own work was regarded by the Church as disrupting the perfect order of the cosmos, Galileo adhered strongly to his belief in such cosmic order—in particular, that all motion was circular. Hence he could not accept Kepler's formulation of elliptical orbits; and this preconception prevented him from giving a full formulation of the inertial law, which he himself discovered, although it is usually attributed to the French mathematician René Descartes. Again, Galileo believed that the inertial path of a body around Earth must be circular. Lacking the idea of Newtonian gravitation, he hoped this would allow him to explain the path of the planets as circular inertial orbits around the sun. Further, the idea of a universal force of gravitation seems to have almost been within Galileo's grasp, but he refused to entertain it because he considered it an "occult" quality.

Thus we see that Galileo combined the rational and empiricist approaches to his research, that he was truly a man of his own time as well as the instigator of the scientific revolution, and that subjectivity colored many parts of his work, both his successes and his failures.

REFERENCES

Biagioli, Mario (1993). *Galileo: Courtier*. Chicago: University of Chicago Press.

Drake, Stillman (1990). *Galileo: Pioneer Scientist*. Toronto: University of Toronto Press.

Einstein, Albert (1967). "Foreword" to Galileo Galilei, *Dialogue Concerning the Two Chief World Systems—Ptolemaic and Copernican*, 2nd ed., translated by Stillman Drake. Berkeley: University of California Press. The foreword is dated July 1952 and is an authorized translation from the German by Sonja Bargmann.

Encyclopaedia Britannica (1975).

Fermi, Laura, and Gilberto Bernardini (1961). *Galileo and the Scientific Revolution.* New York: Basic Books.

Koyre, Alexander A. (1968). *Metaphysics and Measurement.* Cambridge, MA: Harvard University Press.

Mach, Ernst (1960). *The Science of Mechanics.* LaSalle, IL: Open Court.

Miller, Arthur I. (1996). *Insights of Genius.* New York: Springer-Verlag.

Redondi, Pietro (1987). *Galileo: Heretic,* translated by Raymond Rosenthal. Princeton, NJ: Princeton University Press.

Reston, James, Jr. (1994). *Galileo: A Life.* New York: HarperCollins.

Shapere, Dudley (1974). *Galileo: A Philosophical Study.* Chicago: University of Chicago Press.

4.4   WILLIAM HARVEY (1578–1657)

William Harvey was the father of the fields of embryology and physiology and discovered the circulation of the blood.

4.4.A   Brief Biographical Sketch

William Harvey was born on April 1, 1578, in Kent, England. At 16 he entered Cambridge University. It is likely that he began his medical studies there, for the courses offered included practical anatomy. In any event, Harvey received his B.A. degree in 1597 and then went to the medical school at the University of Padua to learn the technical side of medicine. There he studied under Girolamo Fabrizi (1537–1619), also known as Hieronymus Fabricius of Aquadendente, who had demonstrated the valves of the veins and showed that they were always placed so that their open ends faced the heart. Fabricius did not make much of this discovery; it would remain for Harvey to show the importance of the valves for the circulation of blood.

On April 25, 1602, Harvey received his degree as a doctor of medicine and returned to England. He settled in London, but had to be licensed by the College of Physicians to practice in the city. Passing a series of three examinations, but being allowed to practice following his successful completion of the first in 1603, he was finally admitted to the College toward the end of 1604. In the same year he married Elizabeth Browne, the daughter of a prominent physician to the royal family, Lancelot Browne. The marriage was childless but lasted over 40 years. In 1607 he became a Fellow of the College of Physicians, and in 1609 was appointed as a physician to the Hospital of St. Bartholomew. The duties of this post, which included attending at the hospital at least twice a week to examine patients and prescribe for them, formed part of Harvey's workload. In addition, there was his private practice, his work for the College of Physicians—where he was three times censor, charged with inspection of apothecary shops to monitor the quality of the drugs offered for sale—and his experimental activities.

It was in 1616 that Harvey presented his first public anatomical demonstration at the College of Physicians, having been chosen Lumleian Lecturer, a post he held for 40 years. The Lumleian Lecturer was required to give a course of lectures (twice a week for an hour each) on anatomy, surgery, and medicine. The curriculum required a six-year cycle for completion, and the Lumleian Lecturer was also required to carry out a complete anatomical demonstration once a year (on a corpse of an executed criminal).

In 1627, Harvey made an attempt to explain contraction of muscles in *On Local Movement of Animals* and in 1628 he published his magnum opus, *Exericitatio Anatomica de Motu Cordis et Sanguinis in Animalibus (An Anatomical Treatise on the Movement of the Heart and Blood in Animals)*. Here he recounted nine years of thinking and experimentation that convinced him of the circulation of the blood. He spent the bulk of his remaining professional life studying the embryology and gestation characteristics of a wide variety of animal species in an effort to be able to make inferences about how characteristics of humans are passed on from one generation to the next. He referred to this process as the "generation of animals."

In 1631, Harvey was appointed physician-in-ordinary to King Charles I. The two men became friends, and the king made the deer in his deer park available to Harvey for his scientific experiments. When, in 1633, King Charles traveled to Scotland for his coronation as King of Scots, Harvey went along on the trip and used the opportunity to observe the breeding of gannets and other wild birds, thus securing more material for studies of embryology.

By the mid-1630s, Harvey's eminence was sufficiently great that he was called upon to perform such public services as examining women accused of witchcraft, and performing an autopsy on a man reputed to have lived to age 152. In 1639, he became senior physician to the king. During the English Civil War, Harvey was, of course, a royalist. He accompanied the king in his move to Oxford and became affiliated with the university there, receiving an appointment from the king as warden of Merton College when the incumbent was removed for his rebel sympathies. At Oxford, Harvey continued his studies of animal embryology, observing the progress of embryos in a series of chicken eggs. When the rebels besieged Oxford in 1646, Harvey returned to London.

There he published *On the Circulation of Blood*, responding to critics of *De Motu Cordis*. In this work he went further than the original publication in rejecting the anatomical and physiological model dating from the time of Galen (who died in A.D. 200) as to how the body is nourished by the blood. He asserted that the blood gets distributed around the body not in spurts that ebb and peak; rather, the blood flows in a smooth, continuous circulation around the body, and the flow is driven by the pumping and squeezing of the heart.

On his return from Oxford, Harvey found that his lodgings had been broken into and notes and manuscripts, especially those relating to the reproductive life cycle of insects, had been destroyed—a loss of many years of observational work. He was obliged to move to his house in Surrey, because as a Royalist he was not allowed to live in London. It was in Surrey that George Ent insisted on submitting Harvey's manuscript *Exercitationes de Generatione Animalium* for publication, over Harvey's own protests that it was incomplete because of the loss of the insect data. The treatise was published in 1651.

In his later years, Harvey lived with his brother, Eliab, reading rather than investigating new problems and declining on the grounds of age to serve as president of the College of Physicians, although he had been elected to that office in 1654. He died at Eliab's house on June 3, 1657.

4.4.B Harvey's Scientific Contributions

Harvey published two major works, *Exericitatio Anatomica de Motu Cordis et Sanguinis in Animalibus* (*De Motu Cordis*) in 1628 and *Exercitationes de Generatione Animalium* (*De Generatione Animalium*) in 1651. The former is the basis of his lasting fame; the latter has been all but forgotten.

We first briefly discuss *De Generatione Animalium*. This work was based on life-long exercises in dissection and observation, especially of the developing eggs of chickens and of deer, supplied to him in abundance by his patron, King Charles. However careful his observation, Harvey was perforce handicapped by the lack of the existence of a microscope with sufficient resolution to show fine structure, by the rudimentary state of the science of physiology, and by the fact that the development of biochemistry was not to occur until the following century. In this work, Harvey returned to an Aristotelian view of the process of formation of new organisms, insisting on the principle of epigenesis, the formation of a fetus or animal by the addition of one part after another. Indeed, he was the first to apply this term, still current today, to the process. This insistence was contrary to the prevailing view of the time, *preformation*, which held that all parts of the animal were present, if invisible, from the start. Preformation was consistent with the ideas of spontaneous generation, also current at the time (but see Pasteur, Section 4.10). Besides the advocacy of epigenesis, some of the accomplishments of *De Generatione Animalium* include the following (adapted from Pagel, 1967, pp. 329–330): that all animals derive from primordia with the shape and properties of ova. These include true eggs for oviparous animals and the conceptus of mamals (i.e., the fertilized egg). This discovery casts doubt on spontaneous generation; although Harvey did not completely reject the possibility, he narrowed the possible range of application: "Harvey recognized the independent position of the female 'geniture,' the primordial ovum, refuting the Aristotelian view of its passive, merely material, and 'excremental' . . . nature. This applied to all animals, although Harvey failed to recognize the significance of the ovary in mammals. He was fully aware of it in birds, however, and consciously generalized his observations in the latter as applying to the whole animal kingdom" (Pagel, 1967, p. 329). Note that this generalization is a clear example of Harvey's use of subjectivity.

In *De Generatione Animalium*, Harvey also explored the reproductive system, observing peristalsis in the uterus and tubes and the independence of fetal and maternal circulations. He identified amniotic fluid as a source of nutrition, rather than sweat or urine. Starting with a large batch of simultaneously fertilized hen's eggs, Harvey opened one each day and thus constructed a complete description of the development of the chick embryo. Harvey erred, however, in two principal conclusions: that nothing passed from the ovaries to the uterus in mammals, and that the uterus did not receive sperm from the male. Again, we must remember the lack of a microscope and understand with Pagel (1967, p. 328) that the particular mammal to which Harvey had unlimited access, the deer that the king supplied, seems the species least suitable for Harvey's research. The problem with deer is that in this species, the initial stages of pregnancy are much different from what occurs in most other mammals. Deer usually carry but one fetus, and in its early stages such a single fetus would be invisible to the unaided eye. Later, when it first becomes visible, it appears only as an elongated filament some 20 millimeters long and a few millimeters wide. Harvey observed this filament but compared it to a spider thread and did not understand that it served the same function as the sacs that he was used to seeing in other mammals.

Turning to *De Motu Cordis*, we note that to understand Harvey's contribution in describing the circulation of the blood, it is necessary to understand the state of knowledge at the time that he did his work. Singer (1956) gives a detailed history of the ideas of the generation and movement of the blood starting with Galen. Even earlier, Aristotle had declared the heart the most important organ in the body, not only the center of the vascular system but also the source of heat and the seat of intelligence (see Aristotle, Section 4.2). It was common knowledge that the blood was in motion, but the motion was believed to be merely an ebb and flow.

Galen had described the heart as consisting of only two chambers, the ventricles (rather than the four we now know it to have), and maintained that the septum between those chambers was penetrable via invisible passages. Blood was believed to be created in the liver and there charged with natural spirits.

Thence the veins were thought to carry the natural spirits (and nourishment absorbed from the intestines) throughout the body. The veins were also thought to absorb impurities. What Galen called the "arterial vein" (our pulmonary artery) was believed to discharge impurities to the lungs, whence they were exhaled. He also believed that the pulmonary vein (which he called the "venal artery") brought air to the heart and thus accounted for the differing colors of venous and arterial blood—the conception was of two different kinds of blood flowing through two different systems of vessels. Arterial blood was thought to be charged with "vital spirits," accounting for movement and distributed throughout the body. The blood that reached the brain was thought to be there charged with "animal spirits," which were distributed by the nervous system through nerves thought to be hollow.

For centuries the mostly subjectively derived system of Galen had been accepted as true, with anatomy lessons and dissections being used to demonstrate the correctness of his formulations, despite the universal inability to find the pores that were thought to connect the left and right sides of the heart. With the Renaissance, progress began to be made toward a more accurate view. Leonardo da Vinci (1452–1518) was among the first to question Galen's views. Leonardo adopted the empirical approach by dissecting bodies of human beings and animals and found that the bronchi end blindly in the lungs, and he was unable to drive air from the bronchi into the heart. Hence Galen's notion that the pulmonary vein (then called the venal artery) conveyed air from the lungs to the heart was untenable. Leonardo also identified the four chambers of the heart, arguing that all four were of importance; Galen had considered that the only function of the auricles was to prevent overfilling of the ventricles. Leonardo was also able to verify that the valves at the roots of the great arteries functioned to prevent blood from flowing back into the heart.

In the first edition of *The Fabric of the Human Body* (1543), Andreas Vesalius (1514–1564) hinted that Galen might be wrong in maintaining that blood moves from one side of the heart to the other through invisible pores. In the second edition of his work (1555), he made his doubts explicit. Michael Servetus (1511–1553), a theologian, believing that understanding the spirit of man was one route to the comprehension of the spirit of God, discussed physiology in his theological publications. Commenting on Galen, who wrote that throughout the body the arteries and veins mutually give each other blood through openings called *anastomoses*, Servetus insisted that the vital spirit passes from the arteries to the veins through the anastomoses. Thus, by seeing as unidirectional Galen's bidirectional flow, Servetus almost hit on the mechanism of circulation. Servetus says that the many communications between the pulmonary artery and pulmonary vein within the lung confirm this evidence of what is now called the *pulmonary* or *lesser circulation*. Nevertheless, Servetus never took the great step of seeing the pulmonary circulation as connected with the systemic circulation as Harvey was to do; his work was considered heretical, and as a consequence, Servetus was burned at the stake in Geneva in October 1553. This was the milieu that Harvey entered when he embarked on his study of the movement of blood.

Harvey's own teacher, Fabricius of Aquapendente, found the valves on the veins, demonstrating them to the students in his anatomy classes much earlier, but finally publishing his findings in 1600. He did not attach much significance to the valves, failing to inquire deeply about their function but being satisfied that they were merely a means to counteract the effects of gravity and prevent the blood from pooling in the extremities. But Harvey, schooled in the Aristotelian doctrine of final cause (i.e., that everything in nature has a purpose), did wonder about the function of those valves, noting that those in the upper part of the body prevented not downward flow of blood, as would be expected to counteract the effect of gravity but upward flow. Understanding the function of the valves was one of the clues to the system of circulation.

As we shall see, Harvey's work in *De Motu Cordis*, while based on very careful observation, was also hampered by the lack of a proper microscope. His reasoning went approximately as follows: He knew, from the earlier work of Matteo Realdo Colombo (1516–1559), that the active phase of the heart's movement is its contraction, when it become hard and elongated (as a muscle does when it contracts) and is able to expel blood into the aorta. Harvey confirmed this fact and many others by observations on snakes and other cold-blooded creatures whose hearts continue to beat for some time after death and to beat slowly enough to lend themselves to careful and detailed examination. At the same time that the heart contracts, the arteries expand, generating a pulse; that this is a causal chain was confirmed by noting that a cut artery bleeds in spurts coincident with the contraction of the ventricle. That causal chain was extended backward to the auricles; Harvey showed that blood flows from the auricles to the ventricles by cutting a heart open and seeing that at each beat of the auricles there is a spurt of blood into the ventricles.

As Leonardo had shown, blood entering either the aorta or the pulmonary artery is prevented from returning to the heart, and this turns out to be a key idea for Harvey. He calculated that the ventricle can hold about 2 ounces of blood. If the pulse beats 72 times per minute, that means that from the ventricles into the aorta or pulmonary artery must flow about $2 \times 72 \times 60$ ounces (or 540 pounds) of blood in an hour. Harvey wondered where this enormous quantity of blood could come from—the body itself is not large enough to supply it; it cannot come from food and drink, because they cannot be consumed in such quantities; and it cannot lodge in the tissues, for they would burst. Hence, Harvey reasoned, the supply of blood in the ventricles must come from the veins, which also contain blood. Thus the idea of a circular motion seemed to make sense, but now Harvey was faced with answering how the blood gets from the venous system to the left side of the heart and from the arteries to the veins. It was easy to see that blood can enter the right auricle through the vena cava and travel thence to the right ventricle. But the only exit from the right ventricle (if one rules out a route through the septum of the heart, as Harvey did) is through what in Harvey's time was called the arterial vein and what we call the pulmonary artery. Singer (1956, p. 55) quotes Harvey on this point: "The heart then is continually receiving and expelling blood by and from its ventricle, and for this end is equipped with four sets of valves—two for the expulsion of blood and two for its admission. . . . As then the blood is continually flowing into the right side of the heart and flowing out of the left side of the heart, it is obvious that it must somehow pass from the vena cava into the aorta. . . . It thus clearly appears that the blood must penetrate through the porosities of the lung from the right to the left ventricle [to get] from the vena cava to the aorta."

Further, Singer (1956, p. 56) quotes Harvey's conclusion: "I now began to think whether there might not be A MOVEMENT, AS IT WERE, IN A CIRCLE, and this I afterwards found to be true. I saw that the blood, forced by the action of the left ventricle into the arteries, was distributed to the body at large, and its several parts. In the same manner it is sent through the lungs, impelled by the right ventricle into the arterial vein [pulmonary artery]." Harvey then detailed some experiments that clinch the point. In modern terms, using a tourniquet to close off an artery causes blood to back up between the heart and the closure of the artery; closing off a vein, however, causes blood to back up distal to the point of closure.

Arguing that you cannot pass a probe from the trunk of a vein to one of the branches, although it is easy to push it in the opposite direction, Harvey then showed that the function of the valves in the veins is to keep the venous blood flowing to the center of the body from the extremities, without any reverse flow. Finally, Harvey summed up (quoted by Singer, 1956, pp. 62–63):

> Now I may give my view of the circulation of the blood and propose it for general adoption. All things, both argument and ocular demonstration, confirm that the blood passes through lungs and heart by the force of the ventricles, and is driven thence and sent forth to all parts of the body. There it makes its way into the veins and pores of the flesh. It flows by the veins everywhere from the circumference to the centre, from the lesser to the greater veins, and by them is discharged into the vena cava and finally into the right auricle of the heart. [The blood is sent] in such a quantity, in one direction, by the arteries, in the other direction by the veins as cannot possibly be supplied by the ingested food. . . . It is therefore necessary to conclude that the blood in the animals is impelled in a circle, and is in a state of ceaseless movement; this is the act or function of the heart, which it performs by means of its pulse; and that it is the sole and only end of the movement and pulse of the heart.

4.4.C   Major Works

Harvey, William (1627). *De Motu Locali Animalium.*

———(1628). *Exercitatio Anatomica de Motu Cordis et Sanguinis in Animalibus.* Frankfurt.

———(1649). *Execitatio de Circulatione Sanguinis.* Cambridge.

————(1651). *Exercitationes de Generatione Animalium*. London.

————(1766). *Opera Omnia*, 2 vols. London.

Whitteridge, Gweneth (Ed. and Trans.) (1964). *The Anatomical Lectures of William Harvey: Prelectiones Anatomie Universalis; De Musculis*. Edinburgh: E. & S. Livingstone.

Willis, Robert (Trans.) (1847). *The Works of William Harvey, M.D.* London: Sydenham Society.

4.4.D   Subjectivity in the Work of Harvey

Although, as we shall see, Harvey used subjectivity in his work, he was not above decrying it in others. For example, in the preface to *De Generatione Animalium* he writes harshly that his old teacher, Fabricius, "relieth upon probability rather then experience, and layeth aside the verdict of sense which is grounded upon dissections; he flies to petty reasonings borrowed from mechanicks, which is very unbeseeming so famous an Anatomist" (quoted by Keynes, 1966, p. 347). Pagel (1976, p. 3) argues forcefully that Harvey's discovery was not a deduction from a mass of evidence but the result of an idea or a "hunch" that led him to accumulate evidence. He tells us: "[O]ne can find in translations of Harvey a tendency to help the reader to discover in Harvey's text a sequence of events which would today be regarded as legitimate in the process of scientific invention and discovery. On such a view, discovery should normally be the result of a sum total of observations and suitably designed experiments, and such a process should be discernible in Harvey's own account of his discovery as the immediate conclusion from observations and experiments."

But Harvey's own account does not speak of concluding from observations or experiments. We have described what he calls his "meditation" on the quantity of blood issuing from the heart. This meditation or careful introspection led him to the idea of circulation because he reasoned that such a vast quantity of blood could not be manufactured or stored in the body. Harvey actually mentions this meditation twice in the eighth chapter of *De Motu*, both times in the passage in which he announces the discovery of the blood's circulation. He states: "I started to think by myself, whether it (i.e. the blood) could have a motion as it were in a circle. This I afterwards found to be true" (quoted in Pagel, 1976, p. 4).

As an Aristotelian, believing that nature does nothing without a purpose, Harvey was, of course, led to this introspection by his deep knowledge of the anatomy of the heart, veins, and arteries. Harvey also told Robert Boyle that additional considerations of nature's purposiveness led him to postulate the circulation of the blood; keeping arterial blood from flowing back to the heart was the purpose of the valves. Pagel (1976, p. 4) quotes Sir Henry Dale as saying that Harvey's greatest achievement was "not that he made a discovery of such profound and permanent significance, but that he created and displayed for all time the method by which such discovery may be attained and

made secure." Note that the method that Pagel quotes Sir Henry Dale as prais-
ing is one that is based in subjectivity, although confirmed by observation and
experiment.

Another example of the subjectivity in Harvey's work lies in the fact that
he had shown that the blood passes from the arteries to the veins, but had not
been able to physically see the passage. Although the microscope had already
been developed as a by-product of the ideas embedded in the invention of the
telescope (indeed, Galileo demonstrated its use as early as 1610), microscopes
of Harvey's time did not offer sufficiently clear images to be useful in the sort
of biological work that he was doing. Hence he did not use a microscope, and
if he had done so, the power and clarity of the available instruments would
not have been sufficient for him to be able to see the network of capillaries
that we now know connect the arteries with the veins. Wyatt (1924, p. 151)
quotes Harvey's 1649 first *Anatomical Disquisition* to John Riolanus to show
that he tried to find the connections: "I have never succeeded in tracing any
connection between the arteries and the veins by a direct anastomosis of their
orifices; neither in the liver, spleen, lungs, kidneys, or any other viscus is such
a thing to be seen. . . ."

Harvey goes on to describe in detail his thorough but fruitless search for
connections. Having been unable to find the connections but knowing that
blood must flow from arteries to veins, Harvey asserted that the blood must
indeed pass from the arteries to the veins. But he had to rest content with
supposing that the arteries shed their blood into minute spaces in the tissues,
whence it was picked up by the veins. In his 1661 *Anatomical Observations on
the Lungs*, Marcello Malpighi demonstrated the capillary connection.

Yet another example of Harvey's subjectivity in his methodological
approach to his research was described above in a quote from Pagel (1967)
given in Section 4.4.B. There we saw that Harvey had actually seen the ova in
birds, but failing to recognize the significance of the ovaries in mammals, failed
to actually identify a human ovum. Nevertheless, so sure was he of the theoretical concept that was guiding him that he generalized his findings from birds
to the entire animal kingdom.

Still another aspect of subjectivity in the work of Harvey is his adherence
to Aristotelian ideas, even as he proved them false. Aristotle had said that the
heart was the "sovereign" of the body (see Aristotle, Section 4.2). Having just
shown in *De Motu Cordis* that the heart and blood vessels form a single
system, and shown in *De Generatione Animalium* that the blood is formed
before the heart, Harvey nevertheless ends *De Motu Cordis* (as quoted by
Singer, 1956, p. 65) thus:

Are we to agree less with Aristotle in regard to the soveriegnty of the heart? Or,
on the other hand, are we to inquire whether it receives sense and motion from
the brain, and blood from the liver? and whether it be the origin of the veins and
of the blood? They who affirm these latter propositions against Aristotle, overlook,
or do not rightly understand . . . that the heart is the first part which exists, and that

it contains within itself blood, life, sensation, motion, before either the brain or the liver were in being, or had appeared distinctly, or, at all events, before they could perform any function. The heart, ready furnished with its proper organs of motion, like a kind of internal creature, is of a date anterior to the body: first formed nature willed that it should afterwards fashion, nourish, preserve, complete the entire animal, as its work and dwelling-place: the heart, like the prince in a kingdom, in whose hands lie the chief and highest authority, rules over all; it is the original and foundation from which all power is derived, on which all power depends in the animal body.

But it could well have been Aristotle's idea of the primacy of the heart that influenced Harvey to look for an explanation that did not, as Galen's did, require two blood centers, the liver as well as the heart. Aristotle's ideas of unity might well also have encouraged Harvey to think of the unity of the venous and arterial blood rather than of the two systems envisaged by Galen. Further, Artistotle's ideas of the perfection of circular motion could very well have been the wellspring from which Harvey's ideas of circulation flowed.

REFERENCES

Franklin, Kenneth J. (1961). *William Harvey: Englishman*. London: Macgibbon & Kee.

Keynes, Geoffrey (1966). *The Life of William Harvey*. Oxford: Oxford University Press.

Pagel, Walter (1967). *William Harvey's Biological Ideas: Selected Aspects and Historical Background*. New York: Hafner.

———(1976). *New Light on William Harvey*. New York: S. Karger.

Singer, Charles (1956). *The Discovery of the Circulation of the Blood*. London: Wm. Dawson & Sons.

Whitteridge, Gweneth (1967). "William Harvey," pp. 673–675, Vol. 11, *Collier's Encyclopedia*. New York: Crowell Collier and Macmillan.

———(1971). *William Harvey and the Circulation of the Blood*. London: Macdonald; New York: American Elsevier.

Wyatt, R. B. Hervey (1924). *William Harvey*. London: Beonard Parsons; Boston: Small, Maynard.

4.5 SIR ISAAC NEWTON (1642–1727)

Sir Isaac Newton was a mathematician and physicist, who, because of his scientific achievements, is perceived to be, along with Albert Einstein, among the very greatest of all scientists of all time. He created the mathematical theory of the calculus, mathematically codified the universal laws of motion and gravitation, and proposed a physical structure for the propagation of light.

4.5.A Brief Biographical Sketch

Isaac Newton was born prematurely, on Christmas Day, 1642, in England, three months after the death of his father. A tiny and weak baby, Newton was not expected to survive his first day of life, much less 84 years. Within two years his mother remarried and moved to a neighboring village, leaving young Isaac with his grandmother. Thus, Isaac was effectively separated from his mother for nine years until the death of her second husband. They were then reunited for only two years, after which he was sent away to grammar school. The psychotic tendencies that Isaac Newton later sometimes displayed, his anxiety about his work, and the violence of his defense of it have been ascribed to the insecurity engendered by this separation from his mother and his hatred of his stepfather. At school he apparently gained a firm command of Latin, but probably received very little instruction in arithmetic.

Newton was 19 years old when he arrived in Cambridge in 1661, and he was to remain there for 35 years. By this time the scientific revolution was well developed; Copernicus and Kepler had already elaborated the heliocentric system of the universe; and Galileo had proposed the foundations of a new mechanics built on the principle of inertia. But the universities of Europe, including Cambridge, continued to focus on the ideas and teachings of Aristotle, which rested on a geocentric view of the universe, and which dealt with nature in qualitative rather than quantitative terms (see Aristotle, Section 4.2). Therefore, like other college students, Newton began his higher education by immersing himself in Aristotle's work. But even though the new philosophy of the scientific method was not formally in the curriculum, it was becoming common knowledge.

Newton began his studies in mathematics, an area where he felt deficient, because of his minimal earlier preparation. He started with Descartes and went on to classical geometry and the literature of modern analysis, with its application of algebraic techniques to problems of geometry. He mastered the literature in a very short time and then began to move into new territory. He discovered the binomial theorem, and he began informally developing the calculus. Newton made none of this work public, merely keeping his own notebooks. He received the bachelor's degree in April 1665.

Some time during his undergraduate career, Newton discovered the works of the French natural philosopher René Descartes (1596–1650) and the other mechanical philosophers. In contrast to Aristotle, these mechanical philisophers viewed physical reality as composed entirely of particles of matter in motion and held that all the phenomena of nature result from their mechanical interaction. (Newton was eventually to propose that light is composed of particles of matter.) A new set of notes, which he entitled (in Latin) *Certain Philosophical Questions*, begun sometime in 1664, filled the unused pages of a notebook intended for traditional scholastic exercises. Under the title he entered (in Latin) the slogan, "Plato is my friend, Aristotle is my friend, but my best friend is truth." This was clearly the beginning of Newton's scientific career.

In 1665 the Great Plague of London (70,000 deaths in a population estimated at 460,000) closed the university, and for most of the following two years Newton was forced to stay at his home, contemplating at leisure what he had learned. During the plague years Newton laid the foundations of the calculus, developed most of the ideas that would be elaborated in his *Opticks*, and derived the inverse-square relation that the force acting on a planet, attracting it to the Sun, decreases with the square of the planet's distance from the Sun.

Newton was elected to a fellowship in Trinity College, Cambridge, in 1667, after the university reopened. Two years later, at age 26, he was made the

Lucasian professor of mathematics (a position he was to occupy for the next 32 years). The professorship exempted Newton from tutoring but required an annual course of lectures. During the three years 1670–1672, his lectures developed the essay *Of Colours* into a form that was later revised to become Book One of his *Opticks*. His interest in colors led him to the belief that chromatic aberration could never be eliminated from lenses. This problem occurs when undesired colors appear in a lens because the lens acts partially as a prism, as the light is refracted, that is, bent, as it passes through the lens. To study the phenomenon of color, Newton turned to reflecting telescopes, where chromatic aberration would not be a factor because reflection does not involve refraction, and he constructed the first reflecting telescope ever built. Newton's telescope had high power, but because he took advantage of mirroring to increase focal length, it needed to be just 6 inches long. In 1671 the Royal Society of London heard about Newton's telescope and asked to see it. Up until that time his name was probably unknown to the Royal Society (organized just 11 years earlier). He was elected to this distinguished society the following year (1672).

There is no evidence that the theory of colors, fully described by Newton in his inaugural lectures at Cambridge, made any impression, just as there is no evidence that aspects of his mathematics and the content of the *Principia*, 1687 (his most important written work, described later), also pronounced from the podium, made any impression. Rather, the theory of colors, like his later work, was transmitted to the world through the Royal Society of London. Pleased by the Royal Society's enthusiastic reception of the telescope and by his election as a member, Newton volunteered a paper on light and colors early in 1672. On the whole, the paper was also well received, although a few questions and some dissent were heard. Among the most important dissenters to Newton's paper was Robert Hooke, one of the leaders of the Royal Society, who considered himself the master in optics; he wrote a condescending critique of the unknown Newton. The critique would have annoyed a normal man; in Newton it provoked rage and a desire to humiliate Hooke publicly. Less than a year after submitting the paper, he was so unsettled by the give and take of honest discussion that he began to cut his professional ties, and he withdrew into virtual isolation.

In 1675, Newton sent the Royal Society a new paper entitled, "An Hypothesis Explaining the Properties of Light"; it was, in fact, a general system of nature. This paper provoked a new controversy. Hooke apparently claimed that Newton had stolen its content from him, and Newton boiled over again. The issue was quickly controlled, however, by an exchange of formal, excessively polite letters that failed to conceal the complete lack of warmth between the men.

Newton was also engaged in another exchange on his theory of colors with a circle of English Jesuits in Liège, perhaps the most revealing exchange of all. Although their objections were shallow, their contention that his experiments were mistaken lashed him into a fury. The correspondence continued until

1678, when a final shriek of rage from Newton, apparently accompanied by a complete nervous breakdown, was followed by silence. The death of his mother the following year completed his isolation. For six years he withdrew from intellectual commerce except when others initiated a correspondence, which he always broke off as quickly as possible. Starting in the latter part of the 1670s, Newton's alchemical and theological investigations flourished in private, while his public career as a scientist declined.

In 1687, at age 44, Newton published the *Principia*, his magnum opus. The *Principia* was written in Latin and for that reason, was readable by only a small segment of society. The difficulty of the material limited its comprehensibility to an even smaller number. It was composed of the Introduction (including the famous three laws of motion), and three books (Books I and II include discussions of various hypothetical forces and motions, and Book III applies the theory to planetary and terrestrial motions). When the Royal Society received the completed manuscript of Book I in 1686, Hooke claimed that the work was plagiarized, a charge that was difficult to justify. On the other hand, Newton's response to it revealed much about him. Hooke could easily have been mollified by an acknowledgment; he was at that time a sick man who had already passed his prime, and such an acknowledgment would have made Newton appear even bigger. Instead, Newton went through his manuscript and eliminated nearly every reference to Hooke. Such was his fury that he refused either to publish his *Opticks* or to accept the presidency of the Royal Society, until Hooke was dead.

The *Principia* immediately raised Newton to international prominence. In their continuing loyalty to the mechanical ideal, continental scientists rejected Newton's idea of action at a distance for a generation, but even in their rejection they could not withhold their admiration for the technical expertise revealed by the work. Young British scientists spontaneously recognized him as their model. Newton, whose only close contacts with women were his unfulfilled relationship with his mother, who had seemed to abandon him, and his later guardianship of a niece, found satisfaction in the role of patron to a circle of young scientists attracted to him. His four-year friendship with Nicolas Fatio de Duillier (1664–1753), a Swiss-born mathematician resident in London who shared Newton's interests, was probably the most important emotional relationship of Newton's adult life.

Almost immediately following the publication of the *Principia*, Newton, a fervent if unorthodox Protestant, helped to lead the resistance of Cambridge to King James II's attempt to Catholicize it. As a consequence, he was elected to represent the university in the convention that arranged the revolutionary settlement. In this capacity, he made the acquaintance of a broader group, including the philosopher John Locke. Newton tasted the excitement of London life in the aftermath of the *Principia*. The great bulk of his creative work had been completed. He was never again satisfied with the academic cloister, and his desire to change was whetted by Fatio de Duillier's suggestion that he find a position in London. Seek a place he did, and finally, in 1696, at 53, Newton was appointed "warden of the mint," an administrative position within the government that caused him to move to London.

In the meantime, Newton's relations with Fatio de Duillier had undergone a crisis. Fatio de Duillier was taken seriously ill; then family and financial problems threatened to call him home to Switzerland. Newton's distress knew no limits. In 1693 he suggested that Fatio de Duillier move to Cambridge, where Newton would support him, but nothing came of the proposal. Through early 1693 Newton's letters to his friend steadily intensified, and then, without explanation, both their close relationship and their correspondence broke off. Four months later, without prior notice, Samuel Pepys and John Locke, both personal friends of Newton, received wild, accusatory letters. Pepys was informed that Newton would see him no more; Locke was charged with trying to entangle him with women. Both men were alarmed for Newton's sanity; and in fact, Newton had suffered at least his second nervous breakdown. The crisis passed, and Newton recovered his stability. Only briefly did he ever return to sustained scientific work, however, and the move to London was the effective conclusion of his creative activity.

In London, Newton assumed the role of patriarch of English science. In 1703, at age 60, he was elected president of the Royal Society, and he was annually reelected for 25 years. In 1705, Queen Anne knighted him, the first occasion on which a scientist was so honored. Newton ruled the Royal Society magisterially. John Flamsteed, the Astronomer Royal, had occasion to feel that he ruled it tyrannically. In his years at the Royal Observatory at Greenwich, Flamsteed, who was a difficult man in his own right, had collected an unrivaled body of data. Newton had received needed information from him for the *Principia*, and in the 1690s, as he worked on the lunar theory, he again required Flamsteed's data. Annoyed when he could not get all the information he wanted as quickly as he wanted it, Newton assumed a domineering and condescending attitude toward Flamsteed. As president of the Royal Society, he used his influence with the government to be named as chairman of a body of "visitors" responsible for the Royal Observatory; then he tried to force the immediate publication of Flamsteed's catalog of stars. Newton broke agreements that he had made with Flamsteed so that Flamsteed's observations, the fruit of a lifetime of work, were, in effect, seized despite his protests and prepared for the press by his mortal enemy, Edmond Halley. Flamsteed finally won his point and by court order had the printed catalog returned to him before it was generally distributed. He burned the printed sheets, and his assistants brought out an authorized version after his death. In this respect, and at considerable cost to himself, Flamsteed was one of the few men to best Newton. Newton sought his revenge by systematically eliminating references to Flamsteed's help in later editions of the *Principia*.

In Gottfried Wilhelm Leibniz, the German philosopher and mathematician, Newton met his intellectual match. It is now well established that Newton developed the calculus before Leibniz seriously pursued mathematics. It is agreed almost universally that Leibniz later arrived at the calculus independently. There has never been any question that Newton did not publish his *method of fluxions* (his designation for what Leibniz would call *the calculus*); thus, it was Leibniz's paper in 1684 that first made the calculus a matter of public knowledge. In the *Principia* Newton hinted at his method, but he did not really publish it until he appended two papers to the *Opticks* in 1704. By then the priority controversy was already smoldering. If, indeed, it mattered, it would finally be impossible to assess responsibility for the ensuing fracas. What began as mild innuendoes escalated rapidly into blunt charges of plagiarism on both sides. Egged on by followers anxious to win a reputation under his auspices, Newton allowed himself to be drawn into the center of the fray; and once his temper was aroused by accusations of dishonesty, his anger was beyond restraint. Leibniz's conduct of the controversy was not pleasant, yet it paled beside that of Newton. Although he never appeared in public, Newton wrote most of the pieces that appeared in his defense, publishing them under the names of the young men in his entourage. As president of the Royal Society, he appointed an "impartial" committee to investigate the issue, secretly wrote the report officially published by the society, and reviewed it anonymously in the *Philosophical Transactions*. Even Leibniz's death could not allay Newton's wrath, and he continued to pursue the enemy beyond the grave. The battle with Leibniz, the irrepressible need to efface the charge of dishonesty, dominated the final 25 years of Newton's life. It obtruded itself continually upon his consciousness. Almost any paper on any subject from those years is apt to be interrupted by a furious paragraph against the German philosopher, as he honed the instruments of his fury ever more keenly. In the end, only Newton's death ended his wrath. During his last years, his niece, Catherine Barton Conduitt, and her husband lived with him. Newton died on March 20, 1727, at age 84, in London.

4.5.B   Newton's Scientific Contributions

Isaac Newton's major scientific contributions were made principally in mathematics, physics, and astronomy. But he also contributed in such fields as chemistry, alchemy, and theology. In mathematics, physics, and astronomy his contributions were fundamental, codifying, inventive, and enormous in importance and quantity. His research during the three years 1664–1666, ages 21–23, including the period when Cambridge University was closed because of the plague, laid the foundation for all of his subsequent work in science and mathematics. We begin with his more general scientific views and his contributions to chemistry.

The works of the seventeenth-century chemist Robert Boyle provided the foundation for Newton's considerable output in chemistry. Significantly, he had read Henry Moore, the Cambridge Platonist, and was thereby introduced to another intellectual world, the magical Hermetic tradition, which sought to explain natural phenomena in terms of alchemical and magical concepts. (Hermes Trismegistus, the Greek title of the Egyptian moon god, Thoth, was also the god of literature, including all the writings that relate to the sciences. Thoth is also credited with the invention of alchemy and magic.) The two traditions of natural philosophy (i.e., of science), the mechanical and the Hermetic, antithetical though they appear, continued to influence Newton's thought and in their tension supplied the fundamental themes of his scientific career. He was to carry out research in both traditions.

Under the influence of the Hermetic tradition, Newton's conception of nature underwent a decisive change in the late 1670s. Until that time, he had been a mechanical philosopher in the standard seventeenth-century style, explaining natural phenomena by the motions of particles of matter. Thus, he held that the physical reality of light is a stream of tiny corpuscles diverted from its course by the presence of denser or rarer media. His subjective belief was that the apparent attraction of tiny bits of paper to a piece of glass that has been rubbed with cloth is the result of an ethereal flow that streams out of the glass and carries the bits of paper back with it. This mechanical philosophy denied the possibility of action at a distance—as with static electricity, it explained away apparent attractions by means of invisible ethereal mechanisms. Newton's "Hypothesis of Light" of 1675, with its universal ether, was a standard mechanical system of nature. Some phenomena, such as the capacity of chemicals to react only with certain others, puzzled him, however, and he spoke of a "secret principle" by which substances are "sociable" or "unsociable" with others. (There was a similar idea of elective affinity of elements in the work of Lavoisier 100 years later, and of *sociability* of interacting substances in Pasteur's chemistry, 200 years later; see the material on Lavoisier in Section 4.6 and Pasteur in Section 4.10.) About 1679, Newton abandoned the ether and its invisible mechanisms and began to ascribe the puzzling phenomena—chemical affinities, the generation of heat in chemical reactions, surface tension in fluids, capillary action, the cohesion of bodies, and the like—to attractions and repulsions between particles of matter. More than 35 years later, in the second English edition of the *Opticks*, Newton accepted an ether again, although it was an ether that embodied the concept of action at a distance by positing a repulsion between its particles. (Ultimately, the existence of an ether was disproved 200 years later by the Michelson–Morley experiment of 1887; see the discussion in the section on Einstein, Section 4.13.)

The attractions and repulsions of Newton's speculative, and largely subjective, science were direct transpositions of the occult sympathies and antipathies of Hermetic philosophy, as mechanical philosophers never ceased to protest. Newton, however, regarded them as a modification of the mechanical philosophy that rendered it subject to exact mathematical treatment. As he conceived of them, attractions were quantitatively defined, and they offered a bridge to unite the two basic themes of seventeenth-century science: the mechanical tradition, which had dealt primarily with verbal mechanical imagery, and the Pythagorean tradition, which insisted on the mathematical nature of reality. (There are those today who question whether the mathematical modeling of the real world is perhaps too constraining, and that what we really need is to consider approaches that mix the worlds of nature and mathematics; see, e.g., Casti, 1996.) Newton's reconciliation through the concept of "force" (action at a distance), which we now take for granted, may have been his ultimate contribution to science.

Turning to mathematics, we note that Newton is generally considered, along with Archimedes (287–212 B.C.), and Gauss (1777–1855), to have been among the three most important mathematicians of all time. In mathematics, Newton's major contribution was surely the invention of the calculus, but he made other contributions as well. A manuscript dated May 20, 1665 showed that Newton had already developed the calculus to the point where he could compute the tangent and the curvature at any point of a continuous curve. Although Newton published his method only when he appended two papers to *Opticks* in 1704, twenty years earlier, in 1684, Leibniz had already published his own, independent development of the calculus. Thus it was through Leibniz, and not Newton, that the world first became familiar with this new and exciting mathematical methodology. To set down the laws of motion (discussed below), Newton had to define *momentum* (mass times velocity) and to be concerned with rate of change of momentum. To this end he had to define *velocity*, and he defined it as rate of change of position. But what is a rate, and how should it be measured? His need to quantify these terms led him to develop *the differential calculus*. Analogously, his need to quantify the total distance passed over in a given time by a moving particle whose velocity is varying continuously led him to develop *the integral calculus*.

His contributions to algebra included the discovery of a method for finding the rational factors, if any, of a polynomial in one unknown with integral coefficients, and a rule for finding the imaginary roots of such a polynomial; these methods remain in use today. He also did work in infinite series and in number theory. In analytic geometry he introduced polar coordinates, he developed the method of finite differences, and he also worked out a system of classification of curves and developed methods of computation and approximation (e.g., the Newton–Raphson method).

Newton's research in physics focused on optics, both geometric and physical, and the mechanics of bodies in motion (dynamics). His work in optics began very early in his career (1664), when he began experimenting with non-spherical lenses. By 1666, he had concluded that white light is an aggregate of homogeneous colors. He had carried out experiments that involved observing white light pass through a tiny hole in a shutter and then through a glass triangular prism, thus spreading out into a spectrum of colors. Other experiments involved passing light through two prisms. He also studied *interference patterns* resulting from light passing through a convex piece of glass separated by air from a flat piece of glass. The patterns he observed were a series of concentric light- and dark-colored bands. The light bands are formed by the crests of light waves combining to brighten the light, but when the troughs and the crests combine, the light is annihilated, producing darkness. The resulting pattern has since been referred to as *Newton's rings*. In this work, Newton predicated his understanding on a wave theory of light. It was only later that he switched to a corpuscular theory of light.

By 1671, Newton had carried out many experiments with lenses and had so completely convinced himself that chromatic aberration could not be eliminated from ground lenses that he had built the first reflecting telescope, so he could avoid the problem (as discussed in Section 4.5.A). He had presented lectures at Cambridge on optics and the behavior of light (published in 1728 as *Optical Lectures*) and published his first paper on optics in 1672 as "New Theory About Light and Colors" in the *Philosophical Transactions of the Royal Society*.

Newton next began hypothesizing openly in both written talks and in manuscripts about the physical nature of light. Although at first Newton had accepted the prevailing theory of his time that light was a wave motion, the more he experimented with light, the more he became convinced that he had been in error. His later belief was that light is composed of a flow of tiny particles (called the *corpuscular theory* of light). His corpuscular theory stemmed from his study of dynamics of particles and his work on the *Principia*. He came to believe that the behavior of light could be understood by studying how his laws of motion would predict the behavior of the particles comprising the light. In 1704, he published the book *Opticks* (in English), which, among many other things, included his discourses on the interference patterns that we now call Newton's rings, mentioned above. (Interference, as a phenomenon of light behavior, was not actually explained until 100 years later, by Thomas Young.)

A contrary view of the physical nature of light was strongly held by one of Newton's contemporaries, Christiaan Huygens, who believed that light is propagated as a wave motion; this disagreement caused a fundamental intellectual and emotional conflict between Huygens and Newton. Much later, Max Planck (1858–1947) took an intermediate path by proclaiming that light is neither purely particles nor purely waves. It is rather an entity that exhibits both properties under different circumstances: It refracts, reflects, and diffracts, as a wave, but it excites the particles in an atom as a particle and acts as particles in the photoelectric effect. The debate rages on today.

Newton's work in mechanics and dynamics was carried out nearly 80 years after Kepler had described the elliptical orbits of the planets in 1609. Scientists had been trying to work out what it was that kept the planets in their orbits and made the orbits ellipses. It was clear that the Sun had to attract the planets somehow, but what was the attraction, and how did it work? A number of scientists got pretty close to what turned out to be the truth, notably Robert Hooke (1635–1703). When Hooke boasted to Edmond Halley, a famous astronomer and geophysicist who also happened to be secretary to the Royal Society, and who was a great friend of Newton, that he (Hooke) had the answer, Halley went to Newton, in 1684, to check the matter with him. Newton said that he had worked out the answer in 1666 but had never published it.

Halley, in great excitement, urged publication as soon as possible. Newton could do this now with much greater confidence than he could have done it 20 years earlier. For one thing, he now had the calculus, which made easy some calculations that would have been difficult before. For another, he had accurate figures on Earth's size, calculated by John Picard (1620–1682).

Newton took 18 months to write up his work, eventually publishing it in 1687, as *Philosophiae Naturalis Principia Mathematica* (Mathematical Principles of Natural Philosophy; what we call *science* today was called *natural philosophy* in Newton's day), often known simply as the *Principia*. It was written in Latin and did not appear in English until 1729. It is generally considered the greatest science book ever written. Despite the greatness of the book, Newton had trouble publishing it. Hooke was unalterably opposed, and the Royal Society hesitated to become involved in the controversy. Fortunately, Halley had inherited a fortune in 1684, when unknown assailants murdered his father. He saw to the proofreading of the book and had it published at his own expense.

In the introduction to the book, Newton codified Galileo's findings concerning falling bodies (based on Galileo's thought and physical experiments, begun in 1589, and his own experiments) into three laws of motion:

1. The first law of motion enunciated the principle of inertia: *A body at rest remains at rest, and a body in motion remains in motion at a constant velocity (i.e., constant speed in a constant direction) as long as outside forces are not involved*;

2. The second law of motion defines the concept of force. It then asserts that *force is equal to the product of mass and acceleration*. This was the first clear distinction between the mass of a body (representing its resistance to acceleration) and its weight (representing the extent to which it is acted on by a gravitational force).

3. The third law of motion states that *for every action there is an equal and opposite reaction*.

These laws of motion are equivalent to the axioms and postulates with which Euclid began his treatment of geometry. From the axioms and postulates, an enormous number of theorems can be derived, each one building on theorems that went before. In the same way, from the laws of motion, a very large number of mechanical effects can be deduced.

In Book III of the *Principia*, Newton applied the laws of motion to astronomical matters, and thereby, initiated the field of celestial mechanics. He showed, for example, that Kepler's third law relating the periods of the planets (the times they require to rotate around the Sun) to their orbital radii (the distances to their centers of rotation in the elliptical paths they travel) could be deduced from the laws laid out in the introduction to the *Principia*. He was also able to show that the orbital force acting on the Moon followed an inverse-square law. Building on such results, Newton was able to show that the

force of attraction of the Moon to Earth was exactly the same as the acceleration of falling bodies near the surface of Earth. So the force by which the Moon
is retained in its orbit is exactly the same force as we normally call *gravity*.
Newton generalized this result into his *law of universal gravitation*, which
asserts that *all bodies in the universe attract one another with a force
proportional to the product of their masses and inversely proportional to the
squared distance between them*. Newton then went on to calculate the motions
of the tides, of comets, and of the Moon.

4.5.C   Major Works

Newton, Isaac (1687). *Philosophiae Naturalis Principia Mathematica*. Londini: Jussu
     Societaus Regiae ac Typis Josephi Streater.
――――(1729). *Mathematical Principles of Natural Philosophy*. Translated into English
     by Andrew Motte, London: Printed for B. Motte.
――――(1704). *Opticks*. London: Warnock Library.
――――(1707). *Arithmetica Universalis*. Cantabrigiae: Typis Academicus, Londini:
     Impensis Benj. Tooke Bibliopolae juxta Medii Templi Portam in Vico Vulgo
     vocato, Fleet Street.
――――(1720). *Universal Arithmetick*. Translated into English from the Latin by Mr.
     Raphson. London: Printed for J. Senex . . . , W. Taylor . . . , T. Warner . . . , and
     J. Osborn.

4.5.D   Subjectivity in the Work of Newton

Isaac Newton was a strong believer in the experimental method, a procedure
for pursuing scientific research by means of carrying out experiments; this
methodological approach to science was proposed by Galileo (see Section
4.3), an experimentalist who lived a generation earlier. Newton was to become
a supreme experimentalist. His experiments led to hypotheses, and they in turn
led to more experiments. But something special happened on the way from
Galileo to Newton; Newton developed the special mathematical tools to create
mathematical models for the general principles that he believed governed the
phenomena he observed. He was able, through application of the calculus and
other mathematical tools, to quantify the laws of motion precisely. So he could
make predictions and then observe whether his predictions were correct.
Galileo did not have such specialized tools available, so he was less able to
quantitatively verify his beliefs about the laws governing the physical behavior of astronomical and earthly objects.
     Newton became so enamored of his mathematical armamentarium that he
began to believe that through this vehicle he could unlock all the doors that
concealed the mysteries of the universe and that his mathematical physics was
the approach that could solve all problems of the physical world. The more he
labored in his research, the more certain he became that his conclusions
reflected universal truths. His strongly held *prior beliefs* (prior to taking any

data) about what he would find after experimenting on some physical phenomenon became so strong that when there was disagreement between belief and observation, he began to ignore the data and accept the belief. His subjective beliefs began to dominate his attempts to be objective to such an extent that he actually began to alter his data, to force it to conform to his preconceived notions about what he felt his data should have shown. These facts about Newton's subjective approach to his research were revealed only in recent years. Some accounts of his subjectivity follow.

Richard Westfall (1973) suggested that Newton fabricated some of his data to suit the needs of his theory. Westfall carefully examined the three editions of the *Principia* and compared them. His overall conclusion was that Newton had manipulated his data. He noted that while the third edition (1726), the one on which the English translation in use today is based, repeated the second, the second edition (1713) introduced major changes from the original edition (1687) with respect to the three specific major problems of physics that Westfall had decided to investigate: (1) the acceleration of gravity, or the inverse-square law of universal gravitation; (2) the speed of sound; and (3) the precession of the equinoxes. The major changes increased the apparent precision by altering the same data numerically.

In his examination of Newton's reported research on the speed of sound, Westfall makes several allegations of fraud. First, he shows that Newton claims to have calculated the speed of sound and found it to be a value that was identical with the value that Derman had claimed for it, despite the fact that Derman's figure was merely the average of many distinct measurements. Westfall argues that Newton's claimed value for the speed of sound as exactly equal to the *average* value found by Derman was too unlikely a finding to be credible.

Second, Westfall shows Newton claiming that a unit volume of air contained 10% water vapor, and that the water vapor in the air does not vibrate the sound waves that pass through it; that is, sound is not propagated through water vapor. But Newton had not offered any empirical basis for such a claim.

Third, Westfall shows that Newton was claiming that because in his view sound was not propagated through the water vapor in the air, its calculated speed through air (that didn't contain water vapor) should be increased by 10%, the claimed percentage of water vapor in the air (because the speed of sound is slowed down by the water vapor). The 10% figure was only a conjecture and was wrong—the correct percentage varies with temperature and pressure of the air.

Westfall continued his attack on Newton in connection with the *precession of the equinoxes*. The precession of the equinoxes in the plane of Earth's orbit is caused by the cyclic precession of Earth's axis of rotation, that is, the wobbling of Earth's axis caused by the forces of attraction of the Sun and the Moon on Earth's axis. Westfall asserted that because Newton found a lemma to be wrong and had to correct it, he then had to make "an adjustment of more than 50 percent in the remaining numbers. Without even pretending that he had new data, Newton brazenly manipulated the old figures on precession so that he not only covered the apparent discrepancy but carried the demonstration to a higher plane of accuracy."

Roger Cotes was the editor of the second edition of the *Principia*. Westfall points to evidence that he conspired with Newton in the misrepresentations therein as Gale Christianson (1984) explains in another exegesis of Newton's excessive subjectivity:

> As a means of repeatedly underscoring this point, Newton resorted to clever manipulation of his sacred experimental data, feigning a level of mathematical precision that was unattainable by the scientific standards of his time.
>
> In point of fact, Newton took relatively little risk by doctoring his figures, as he was well aware. Unlike the individual who intentionally fakes his experimental data to confirm a promising hypothesis, he knew full well that the fundamental insights gathered from a lifetime of meticulous experimentation were sound. He had simply, if dishonestly, elevated these demonstrations to a higher plane of accuracy that the figures warranted. When viewed in this light, such pretense obviously had one overriding purpose: to serve as a grand polemic against the mechanists in general and against Leibniz in particular. Writing Newton on February 11, 1712, concerning the problem of calculating the rate of the moon's descent, Cotes captured the spirit of the enterprise: "In the Scholium to the IVth Proposition I think the length of the Pendulum should not be put 3 feet & 8 2/5 lines [a line is 1/12 of an inch]; for the descent will then be 15 feet 1 inch 11/3 line. I have considered how to make that Scholium appear to the best advantage as to the numbers and I propose to alter it thus." This was but one of several instances in which the young mathematician proved himself almost as adroit as his master at doctoring the figures. Two months later Newton praised him for further manipulating the lunar theory.

It seems clear that Newton was so certain of the truth of his mathematical and physical results that he was prepared to modify his data to silence his opponents. He appears to have fudged his data to demonstrate how accurately his theory had been confirmed by evidence, when in fact, his theory had not yet been confirmed, except to his own satisfaction.

Simon Schaffer (1989) has commented on Newton's subjectivity in his numerous experiments. He points out that Newton has been accused of various crimes of improper experimental procedure in connection with his extensive experiments in geometric optics. He has been accused of not taking a sufficiently large sample size to prove his claims, and he has been accused of reporting experiments that cannot be replicated by others, a minimal requirement for acceptable scientific procedure. Then Schaffer (1989, p. 68) quotes Newton's own words in his first published account of the optical trials: "that 'the historical narration of these experiments would make a discourse too tedious and confused, and therefore lay down the *Doctrine* first and then, for its examination, give you an instance or two of the *Experiments*, as a specimen of the rest.'" Thus Schaffer is also pointing out that Newton was criticized for advocating stating one's principle first, and then illustrating it with a couple of examples, as the appropriate way to do science. In fact, in some respects, Newton's subjectivity overwhelmed his experimental procedure. Moreover, his need to convince his peers of the validity of the research results he subjectively believed in very strongly drove him to report numerical results that strongly and unquestioningly corroborated his predictions.

REFERENCES

Asimov, Isaac (1989). *Asimov's Chronology of Science and Discovery*. New York: Harper & Row.

Bell, E. T. (1937). *Men of Mathematics*. New York: Simon & Schuster.

Casti, John (1996). "Confronting Science's Logical Limits," *Scientific American*, October, pp. 102–105.

Christianson, Gale E. (1984). *In the Presence of the Creator: Isaac Newton and His Times*. New York: Free Press.

Encyclopaedia Britannica (1975). Vol. 13.

Encyclopedia Americana (1958). Vol. 20.

Fauvel, John, Raymond Flood, Michael Shortlans, and Robin Wilson (Eds.) (1988). *Let Newton Be! A New Perspective on His Life and Works*. Oxford: Oxford University Press.

Gillispie, Charles Coulston (Ed.) (1974). *Dictionary of Scientific Biography*, Vol. 10. New York: Charles Scribner's Sons.

Schaffer, Simon (1989). "Glass Works: Newton's Prisms and the Uses of Experiment," in David Gooding, Trevor Pinch, and Simon Schaffer (Eds.), *The Uses of Experiment: Studies in the Natural Sciences*. Cambridge: Cambridge University Press.

Westfall, Richard (1973). "Newton and the Fudge Factor," *Science*, **179**:751–758.

4.6 ANTOINE LAVOISIER (1743–1794)

Antoine Lavoisier has been called the father of modern chemistry. He disproved the phlogiston theory of combustion, correctly identified the role of oxygen in combustion and other forms of oxidation, and provided a systematic classification of chemical substances and a corresponding nomenclature.

4.6.A Brief Biographical Sketch

Antoine-Laurent Lavoisier was born in Paris on August 23, 1743. After the death of his mother in 1748, Lavoisier (together with a younger sister who died as a teenager) was cared for by his father, a lawyer, and a maiden aunt, his mother's sister. From age 11 he was educated at the College Mazarin, receiving training in mathematics and the sciences as well as in language, literature, and philosophy. At age 18 he transferred to the Faculty of Law and received his baccalaureate in law in 1763 and his licentiate in 1764. But his main interests, even while studying the law, were in the sciences, and he soon abandoned the law in order to devote himself to science. He was particularly interested in geology and set out to learn the chemistry he would need to identify and classify rocks. He studied chemistry with Guillaume François Rouelle, who is credited with introducing two generations of Frenchmen to the subject.

For several years Lavoisier traveled with a family friend, Jean-Etienne Guettard, collecting rocks and minerals and preparing a geologic atlas of northern France. During these expeditions Lavoisier displayed the strong interest in quantification that was to mark all his work as a scientist and did preliminary work on a theory of stratification that was finally to be presented to the Royal Academy of Sciences in 1788. Lavoisier's first chemical research grew out of this interest in geology. He began to investigate the properties of the mineral gypsum (a hydrated sulfate of calcium). He measured the solubility of different samples of gypsum and of what was called calcinated gypsum (plaster of paris) and presented these papers to the Royal Academy in 1765 and 1766. He was eager for election to the Academy, not for financial reasons, because his own circumstances were comfortable (indeed, even before he reached the age of majority, his father made over to him a sizable inheritance). The election was important, nevertheless, so that he would have a prestigious institutional base from which to carry out his scientific studies. To contribute to the goal of election, Lavoisier entered an Academy-sponsored competition to design street lighting for Paris. He carried out an extensive theoretical and practical study, an early example of Lavoisier's penchant for combining theoretical and practical work, particularly in an effort to carry out projects for public service. Although he did not win the offered prize (which was split among several contestants who had concentrated on the practical aspects of the problem), his work was much admired and the Academy awarded him a gold medal for it. Nevertheless, he was not elected to the Academy to fill the vacancy that occurred in 1766.

By 1768, however, Lavoisier had read two more papers before the Royal Academy, these concerning the composition of water and also growing out of his interest in geology. In the election to fill another vacancy that had occurred in the Academy, he received a slim majority of the votes; King Louis XVI, who had the privilege of making the final choice between two candidates recommended by the Academy, failed to choose Lavoisier. Nevertheless, the Academy bent its rules slightly and admitted Lavoisier as an extra member. From that time forward, Lavoisier, in addition to his own research, was active in the investigations of the Academy until its dissolution after the French Revolution. He served on committees (and often served as their secretary) to investigate such issues as hypnotism, hydrogen balloons, and the conditions of prisons and hospitals in Paris.

In 1768, Lavoisier made another decision that was to prove momentous. He used part of his inheritance to buy a part of a share in the *Ferme Général*. The *Ferme* was an organization of businessmen who entered into a contract with the government to collect taxes on tobacco, salt, and produce entering Paris, and to collect customs duties. Although all evidence seems to point to Lavoisier's honest business dealings, it would seem that the contracting out tax collection, with the contractors paying a fixed sum to the government and being permitted to pocket any extra revenue collected, was fated to create resentment among those from whom taxes were being collected. In pre-Revolutionary France, taxes were collected only from the middle and working classes, the nobility being exempt. Thus, eventually the *Fermiers Générales* would not be in good repute with the revolutionaries.

Despite his travels on behalf of the *Ferme Générale*, Lavoisier continued his research program, often reporting on his results to scientific bodies in the cities he visited. Another circumstance that was to put Lavoisier in peril during the Reign of Terror was the purchase by his father, in 1775, of an honorific office that carried a title of hereditary nobility. Lavoisier inherited that title and also purchased a country estate. Despite the fact that he used the estate to carry out agricultural experiments and reported the results to the Royal Agricultural Society of Paris (of which he was an elected member) and to the government's Committee on Agriculture, his possession of it made him part of the landed aristocracy, a status that was to be considered inimical to the Revolution.

Lavoisier's concern with the composition of waters from various parts of France led to his interest in a plan to bring water to Paris in an open canal and the means of testing such water for potability. The standard test was to evaporate the water to dryness and then weigh the residue. But at that time it was believed that there were just the four basic elements described by Aristotle: earth, air, fire, and water. Many scientists believed that distilling water could transmute it to earth. If such transmutation indeed took place, the evaporation method would be unreliable. Lavoisier devised an experiment in which he kept water that he had carefully distilled at the boiling point for 101 days. Having weighed the flask beforehand, Lavoisier found that it had lost weight after the passage of three months' time. When he evaporated the remaining water to dryness and weighed the residue together with the solid particles that he found appearing in the water during the prolonged heating, these accounted roughly for the weight lost by the flask. It was easy for Lavoisier to conclude that the appearance of solid material was not the result of transmutation but of the dissolving of the glass of the flask. This work, initialed by the secretary of the Academy as being completed in 1769 and presented in late 1770, was an early step in Lavoisier's path to investigating the elements and how they combine with one another.

In 1771, at the age of 28, Lavoisier married the 14-year-old Marie Anne Pierrette Paulze, daughter of a fellow *Fermier Générale*, Jacques Paulze. By all accounts the marriage was a very happy one, although childless. Madame Lavoisier learned English so that she could translate the scientific literature in that language for her husband's use. She also studied drawing so that she could illustrate his scientific work.

In 1775, at age 32, Lavoisier was appointed a commissioner of the Royal Gunpowder Administration. His duties were essentially those of a scientific director, and a residence at the Paris Arsenal was part of the appointment.

Here for the next two decades, in a laboratory that he himself carefully (and expensively) equipped, Lavoisier was to carry out his major innovative scientific work. From here he conducted the experiments that convinced him of the falsity of the phlogiston theory of combustion; from here he formulated his new vision of chemistry. To persuade other chemists, Lavoisier worked with Claude Berthollet, Antoine de Fourcroy, and Guyton de Morveau to publish a volume proposing a new system of chemical nomenclature that was consistent with the new theory. In 1789, at the start of the French Revolution, for the same purpose of persuasion, Lavoisier published his classic text, *Traité élementaire de chimie.*

Lavoisier, a political liberal, took an active part in the events leading up to the French Revolution. He served as an alternative deputy for the nobility of Blois to the Estates General and there drafted a proposal to redress grievances. Among other services, he was a member of the Bureau of Arts and Crafts, for which he prepared a report on public instruction that proposed a system of universal free education and outlined a curriculum. As the Revolution moved to the left, Lavoisier continued in his scientific pursuits as much as was possible. Indeed, in 1791 he served on a committee of the Academy to develop a uniform system of weights and measures for the nation. But the institutions in which Lavoisier had served were being abolished by the Revolution. In 1791, the *Ferme Générale* was abolished. In the same year he was removed from his post in the Gunpowder Administration. In 1793 he had to move from his residence at the arsenal; and in the same year, the beginning of the Reign of Terror, all the royal learned societies, including the Academy of Sciences, were abolished.

On November 28, 1793, Lavoisier was arrested. We can discern several reasons that predisposed him to be a victim of the Terror. First, he had been a member of the *Ferme Générale,* and it had been his proposal, years earlier, that a wall be built around Paris to prevent smuggling of produce into the city. The theory was that such a wall would protect honest traders, and his proposal had been accepted and carried out. Not only was the expense of building a needlessly elaborate wall outrageously high and hence strongly resented, but it was widely believed to function to restrict air circulation, confining bad air to the city and preventing good air from entering. There had also been a dangerous incident in 1789 in which the revolutionary mob had seized some low-grade industrial powder that was being shipped abroad and, in the name of liberating it for the Revolution, brought it for storage at the arsenal where Lavoisier was powder commissioner. He realized that the low-grade powder was occupying storage space needed for musket powder, and hence sought and received a permit to ship the industrial powder elsewhere, to be replaced by musket powder. Through a mix-up in the paperwork, the citizens of Paris believed that the powder they had liberated for the Revolution was being disposed of traitorously by Lavoisier. Although the situation was eventually clarified, some hint of suspicion clung to Lavoisier. Lavoisier had also, in the early 1780s, made an enemy of Jean Paul Marat, a would-be chemist and a revolutionary zealot, by stating publicly that a paper Marat had written entitled "*Recherches physiques de la feu*" was devoid of merit and would never be approved by the Academy for publication. (Lavoisier had been irate that the work, on the basis of no solid empirical evidence, presented a theory of combustion that challenged his own; he was further outraged when a notice in a Paris journal stated falsely that the paper had been approved by the Academy.) Marat was a powerful enemy. As early as January 1791 his newspaper *L'ami du peuple* had launched a first attack on Lavoisier.

The specific charge against Lavoisier and 31 of his colleagues (including his father-in-law) arrested at the same time was that they had defrauded the nation of 400,000,000 livres. Although that amount was later reduced, all the accused were kept imprisoned, their property confiscated, and their homes sold by the state, and it was determined that they be tried by the revolutionary tribunal as enemies of the state. The deputy in charge of the investigation, André Dupin, was approached by friends of Lavoisier and convinced to separate Lavoisier's case from that of the other Farmers and to eliminate from the final report all accusations that might condemn him. Dupin made the condition, however, that Madame Lavoisier make the request to him in person. Madame Lavoisier did indeed visit Dupin, but instead of pleading for her husband's life and thus mollifying the petty official, she haughtily asserted his innocence and that of the other Farmers. She argued that he would be disgraced if his case were separated from those of the other Farmers and demanded that he be set free. Thus Lavoisier's fate was sealed. Dupin presented his report calling for a trial before the revolutionary tribunal on May 5, 1794. The trial began at 10 in the morning of May 8 after the accused had been permitted 15-minute meetings with their court-appointed defenders. The accused were found guilty of "a plot against the people tending to favor the enemies of France by excising, extorting, exacting from the people by adding water to tobacco . . . by retarding the warfare of the nation against despots who rise against the Republic." Lavoisier and his fellow Farmers were condemned to death within 48 hours, but they were guillotined that same afternoon.

4.6.B Lavoisier's Scientific Contributions

To understand Lavoisier's contributions, it is necessary to review the state of chemical knowledge and theory at the middle of the eighteenth century. The following account is adapted from McKenzie (1960).

The theory of four elements of the Greeks (earth, air, fire, and water), remained the overarching conceptualization of fundamental chemical substances. The specific explanation for combustion was embodied in the *phlogiston* theory. All flammable substances were thought to contain this mysterious substance, which is given off on burning. The amount of phlogiston in a substance was thought to be measured by the amount of physical residue left after combustion. The less physical residue left, the higher the proportion of phlogiston the substance contained.

Because charcoal was used in refining metal ore and disappears as the ore forms, it was thought that metals were combinations of ore and charcoal. Therefore, metals must contain the phlogiston absorbed from the charcoal. But some metals when heated to molten produce a surface scum, called a *calx*, in a process called *calcination*. The calx was thought to be the metal after the loss of phlogiston. That the calx weighed more than the metal was puzzling—if the metal incorporated both phlogiston and calx, how could it weigh less than the calx itself? This puzzle was solved (probably to nobody's satisfaction, certainly not to Lavoisier's) by attributing *negative weight* to phlogiston. Another explanation, advanced by Robert Boyle, was that in the burning process, fire particles enter the calx.

It was also known that combustion (and calcination) could only take place in air—never in a vacuum—and persists only briefly in limited air. Phlogiston theory explained this phenomenon by postulating that the phlogiston given up by combustion was absorbed by the air and that air has limited capacity to absorb phlogiston. Air was the only gas thought to exist. Joseph Black, in 1756, in a major breakthrough, identified another gas, which he termed *fixed air* and which we know as carbon dioxide.

Although there seems to be no evidence that Lavoisier was interested in combustion, the calcination of metals, or the composition of air prior to 1772, on February 20, 1773 he recorded in his laboratory notebook his intention of embarking upon a large series of experiments on the elastic fluids (gases) that bodies emit during chemical reactions and on the air absorbed during combustion. He visualized that his results would be important enough to bring about a revolution in physics and chemistry. Guerlac (1973, p. 73) characterizes this revolution:

> This revolution has often been summed up as Lavoisier's overthrow of the phlogiston theory (his new chemistry was later called the antiphlogistic chemistry), but this is only part of the story. His eventual recognition that the atmosphere is composed of different gases that take part in chemical reactions was followed by his demonstration that a particular kind of air, oxygen gas, is the agent active in combustion and calcination. Once the role of oxygen was understood—it had been prepared before Lavoisier first by Scheele in Sweden and then independently by Priestley in England—the composition of many substances, notably the oxyacids, could be precisely determined by Lavoisier and his disciples. But the discovery of the role of oxygen was not sufficient of itself to justify abandoning the phlogiston theory of combustion. To explain this process, Lavoisier had to account for the production of heat and light when substances burn. It is here that Lavoisier's theory of the gaseous state, and what for a time was his theory of the elements, came to play a central part.

During 1772, a year which Guerlac (1961) calls crucial for Lavoisier and for chemistry, Lavoisier had carried out some important experiments. On November 2, 1772 he deposited a note with the Secretary of the Academy stating that he had been able to show that when phosphorus and sulfur are burned, they gain in weight because of the air that is fixed during the burning. Further, he stated that the increase in weight could occur in all bodies that gain weight by combustion and calcination. He had confirmed that latter speculation in the case of lead by reducing lead oxide in a closed vessel and observed that air was given off when the calx changed into the metal. In the succeeding year Lavoisier undertook to repeat others' experiments and to extend his own. By the Easter meeting of the Academy, Lavoisier presented what has become a most famous paper, entitled "On a new theory of the calcination of metals." Here he pointed out that when exposed to air, iron rusts and copper is reduced to verdigris (hydrous copper oxide). As these metals pass from the metallic state to the powder, their weight increases. His experiments led him to conclude that (as quoted by Donovan, 1993, p. 750) "there is an absorption of air during calcination and a disengagement of air during reduction" and therefore, "(1) a metallic powder is formed by the combination of a metal with a certain quantity of fixed air, and (2) metallic reduction consists essentially of the disengagement of the air with which the metal has in some manner been saturated." Because the abundant fixed air in the atmosphere causes the gain of weight in calcination, Lavoisier saw no need to consider whether phlogiston is also involved. These findings were written up in Lavoisier's first book, *Opuscules physiques et chimiques.*

Later, in 1774, Lavoisier read a paper to the Academy on the calcination of metals in sealed vessels. Boyle's experiment that had caused him to conclude that particles of fire had entered the tin was arranged in such a way that he heated the tin in a sealed container, cooled it, unsealed it, and then weighed it. Lavoisier, on the other hand, weighed the vessel after heating but before unsealing and discovered that the weight remained the same, lending evidence to his contention that the gain in weight of the tin on calcination was due to the absorption of something from the air within the sealed vessel. Further, when he unsealed the vessel, he heard a whistling noise as air rushed in and determined that the weight of the additional air was equal to the weight added to the tin. Although he was promptly informed that similar experiments had been carried out earlier by a Father Beccaria and that Jean Rey had postulated in 1630 that the weight gain in calcination was due to the absorption of air, Lavoisier persevered. It was not the experimental findings as such for which we remember him, but their weaving into a coherent theory.

It had become clear to Lavoisier that calcination in a closed system does not continue until all the air is absorbed, but ceases when some fraction is exhausted. But the reason for this phenomenon eluded him until he learned of findings of Joseph Priestley, who had heated red calx of mercury (mercuric oxide) and found that the calx reverted to mercury and a gas was driven off. Priestley found that this gas supported combustion better than did normal air. He rejected the idea that it could be a gas with properties similar to those that he had recently discovered in nitric oxide. His reasoning was that calx or mercury does not contain what was called "nitre" and thus could not produce nitric oxide. Priestley further discovered that the new gas was eminently respirable, using both mice and himself to test its "goodness." Although Priestley called the new gas *dephlogisticated air* because its support of combustion and respiration suggested that it was able to absorb great amounts of phlogiston, Lavoisier was later to christen it with the name we know today, *oxygen*. (The name connotes "begetter of acids," as Lavoisier, in his effort to classify and systematically name all chemical substances, mistakenly believed that all acids contained oxygen.) Priestley visited Paris in late 1774 and told many there, including Lavoisier, about his new gas.

Lavoisier repeated Priestley's experiments and in April 1775 read to the Academy a paper entitled "On the Nature of the Principle that combines with the Metals on Calcination and increases their Weight." In this paper he described two experiments with calx of mercury. When he heated it together with charcoal, a gas was given off that did not support either combustion or respiration and that precipitated lime water (calcium hydroxide). These tests showed that it was fixed air (carbon dioxide). If heated alone, however, calx of mercury gave off a different gas, one that did support both combustion and respiration but did not precipitate lime water. From all this Lavoisier concluded that what combines with metals is particularly pure air; the notion that air is a mixture rather than an element had yet to be discovered.

It was in 1777–1778 that the next part of the puzzle fell into place, as a result of a crucial experiment. Four ounces of mercury were heated until calx ceased to form in an open container that was set over a bath of mercury, the whole covered by a glass container called a bell jar. The volume of air in the bell jar dropped from 50 to 42 cubic inches and those 42 cubic inches of gas could not support combustion or respiration. When the deposited calx was heated to a yet higher temperature, it gave off 8 cubic inches of air, which exhibited all the properties of dephlogisticated air. When the 8 cubic inches of dephlogisticated air were mixed with the 42 cubic inches of air remaining from the first procedure, 50 cubic inches of normal air were recovered. There was no need, or room, for phlogiston in this explanation.

Further support for Lavoisier's theory that combustion involves combination with oxygen arose from his interpretation of experiments carried out by Henry Cavendish and by Priestley. Cavendish had isolated a gas which he called *inflammable air* becausd it burned in air, and which he regarded as almost pure phlogiston. In 1781, Priestley used an electric spark to explode a mixture of inflammable air and dephlogisticated air and found that a dew was produced. In the same year, Cavendish undertook a series of experiments with inflammable air but did not publish his results until three years later. Exploding common air and inflammable air, he again produced dew, the dew this time accounting for all the inflammable air and one-fifth of the common air. His analysis of the dew suggested strongly that it was pure water. From these experiments Cavendish concluded that inflammable air was more likely to be a mixture of water and phlogiston than pure phlogiston, as it needs a flame to react with dephlogisticated air.

When Cavendish's assistant visited Paris in 1783, he informed Lavoisier of these unpublished experiments. According to the theory that Lavoisier was developing at the time, the product of the burning of inflammable air in the air should have been an acid. Impressed, he repeated Cavendish's experiments in a very rough form and reported the results to the Academy with barely a mention of Cavendish. (The results reported included detailed calculations that showed that the sum of the weights of the two gases was equal to the weight of the water produced—calculations that were too exact to have been derived from his own experiment and must have arisen from the work of Cavendish.) Lavoisier never replied to the charge of plagiarism arising from this incident.

The genius of Lavoisier lay in his interpretation of these results. He immediately saw that inflammable air (which he was to label *hydrogen*, meaning "water former") and oxygen were elements that combined into a new substance, a chemical compound, water. He also reversed the experimental program of Cavendish and performed an analysis of water by dropping it through the red-hot iron barrel of a gun. The oxygen combined with the iron to form an iron oxide and the hydrogen was released.

To Lavoisier's mind the revolution in chemistry that he had promised in his 1772 memorandum (the demonstration that on burning, all substances gain weight) had been achieved; his remaining task was to see to its acceptance. For this purpose he gathered the group of collaborators: Berthollet, de Fourcroy, and de Morveau. He felt that the names of substances should reflect their chemical composition or their actions and should use Latin and Greek roots. As we have already seen, inflammable air, because it was involved in the synthesis of water, was named hydrogen and dephlogisticated air was named oxygen because Lavoisier thought it formed acids. Calces of metals were called *oxides*, the main ingredient of charcoal was called *carbon*, fixed air was called *carbonic acid*, and its salts were called *carbonates*. Vitriolic acid, formed by the combination of sulfur and oxygen, was called sulfuric acid and its salts, sulfates; Vitriol of Venus was called *copper sulfate*. To further the acceptance of both his theory and his system of nomenclature, Lavoisier undertook to write a text, which was published in 1789 and entitled *Traité Elementaire de Chimie*. In the text, Lavoisier defined an *element* as a substance that cannot be decomposed into anything simpler (or at least had not yet been successfully so decomposed). He gave a list of elements; they were 33 in number and included 23 that we accept as elements today. The list also includes light and *caloric*, which Lavoisier regarded as the principle of heat. He explained that matter may exist in any of the three states solid, liquid, or gas. But he theorized that the state of a substance depended on the amount of the imponderable caloric associated with the substance. Although the work of Cavendish and Priestley as well as that of Lavoisier himself had tacitly assumed a law of conservation of mass, in that all these scientists felt constrained to show that the mass of the products of a chemical reaction was the same as the mass of the starting substances, that law had never been explicitly enunciated. The text did so.

Holmes (1985) entitles his book-length biography *Lavoisier and the Chemistry of Life* and concentrates on Lavoisier's contributions to physiological chemistry. He states (p. xv): "He devised the basic method for the elementary analysis of plant and animal matter around which the field of organic chemistry afterward coalesced. His famous study of the alcoholic fermentation of sugar was the starting point for investigations of that process which have occupied physiologists and chemists ever since. His theory that respiration is a slow combustion of carbon and hydrogen has been central to the whole of modern biology."

4.6.C   Major Works

A bibliography of Lavoisier's works was produced by Denis I. Duveen and Herbert S. Klickstein (1954), *A Bibliography of the Works of Antoine Laurent Lavoisier, 1743–1794*, London; W. Dawson & Sons, and E. Weil. *Oeuvres de Lavoisier*, 6 vols., Paris, 1862–1893, contains his published articles, some unpublished material, and two of his major books.

Lavoisier, A. L. (1774). *Opuscules Physiques et Chimiques*. Paris. (An English translation by Thomas Henry entitled *Essays Physical and Chemical* was published in London in 1776.)

———(1787). *Méthode de Nomeclature Chimique, Proposee par MM. de Morveau, Lavoisier, Bertholet [sic] & de Fourcroy*. On y a joint un nouveau système de caractères chimiques, adaptés a cette nomenclature, par MM. Hassenfratz & Adet. Paris. (An English translation by James St. John was published in London in 1788.)

———(1789). *Traité Élementaire de Chimie*. Paris. (An English translation by Robert Kerr entitled *Elements of Chemistry, in a New Systematic Order, Containing All the Modern Discoveries* was published in Edinburgh in 1790.)

———(1805). *Mémoires de Chimie*, 2 vols. Printed posthumously for Madame Lavoisier from incomplete proofs partially corrected by Lavoisier, not commercially published.

4.6.D   Subjectivity in the Work of Lavoisier

Because a large proportion of Lavoisier's contribution to science consists of a theoretical synthesis of experimental results, many of them originally obtained by other researchers, there is a sense in which we can consider his whole body of work as the product of informal subjectivity. (Such informal subjectivity may be compared with the formal subjectivity considered in Chapter 5.)

In another broad sense of subjectivity, Donovan (1993, p. 79) tells how Lavoisier's mind was prepared to accept different conceptualizations of elements. He refers to Lavoisier's teacher, Guillaume François Rouelle, and says:

Like Rouelle, Lavoisier treated the identification and characterization of the chemical elements as a matter of investigative tactics rather than philosophical ontology. He posited the existence of simple substances because he had to have some "things" to reason with, not because he believed he could gain absolutely secure knowledge about the ultimate units of matter. Earth, air, fire, and water provided reasonable starting points for physicists and chemists, yet Lavoisier was prepared to consider the possibility that water, for instance, can be turned into earth or air. The kind of knowledge science builds on is experimental and quantitative. The simplest substances that can be known are those that resist all further attempts at analysis. In principle, therefore, Lavoisier was prepared from the outset to accept the plurality of airs and the analysis of water into oxygen and hydrogen. He had learned from Rouelle to treat theories of elements heuristically. When formulating his own theories, he was fully aware that the "things" he called simple substances were historically and instrumentally contingent.

Thus Donovan shows us the flexibility of Lavoisier's subjectivity—if his current instruments do not let him decompose a substance, he will consider it an element *pro tem*, with the classification being subject to possible future amendment when instruments are improved.

In a more specific sense, Lavoisier early on claimed to have rejected the use of subjectivity in his research, despite what we shall see is a generous subjective leavening in his objective procedures. In his earliest paper given to the Academy of Sciences (on the analysis of gypsum) he insisted that in reporting experiments a scientist must give only demonstrated facts and must avoid speculations. Thus, he claimed that although he thought he could guess the cause of the phenomenon that overheated gypsum would not recombine with water, he was unable to prove his case and so he refused to speculate. As early as 1764 he drafted an outline for a course in chemistry. There he cited Benjamin Franklin's experiment in which he captured lightning and stored it in a Leyden jar as changing the identification of lightning and electricity from an analogy to a demonstration. Indeed, he said there, and his career bears out, that what he valued in an experiment was not its ability to reveal surprising facts but its provision of demonstrative evidence. Thus, Lavoisier insisted on the importance of experiment and derides reliance on hypotheses. As we shall see, Lavoisier did not always practice what he preached.

Lavoisier was a man of his time, and the received wisdom of his time was that combustion caused a substance to lose phlogiston. Although Lavoisier's program of research was to overthrow the phlogiston theory (and indeed became known as the antiphlogiston theory), he used explanations derived from phlogiston theory, at least early in his career, to interpret his experimental results. For example, in 1772 he found experimentally that phosphorus absorbs air in burning and increases in weight. His published explanation of these results was that the phosphorus was decomposed and lost phlogiston. Note that this explanation would have to postulate a negative weight for the mysterious and elusive phlogiston. In his masterly summation of his work on chemistry, the *Traité élementaire de chimie*, while clearly proving the lack of necessity of the imponderable phlogiston to explain combustion, he invoked the equally imponderable caloric to explain the differing states of matter. Ironically, it was Benjamin Thompson, Count Rumford, who both married Lavoisier's widow and, by showing that heat is a form of energy, proved that Lavoisier's caloric was as unnecessary a construct as phlogiston had been.

In 1773, writing in his laboratory notebook (and quoted by Guerlac, 1961, p. 228), Lavoisier maintained: "I have felt bound to look upon all that has been done before me as merely suggestive; I have proposed to repeat it all with new safeguards, in order to link our knowledge of the air that goes into combustion or that is liberated from substances, with other acquired knowledge to form a theory."

There is some question whether Lavoisier did, in fact, repeat the experiments of others (we have seen above that there were accusations that Lavoisier plagiarized the work of both Priestley and Cavendish), let alone whether they were repeated "with new safeguards." But experimental findings that supported his emerging new paradigm must have seemed appealingly accurate to Lavoisier. Indeed, Davis (1966) tells us that some of the readers (including Priestley) of Lavoisier's first book, *Opscules physiques et chimiques*, found that his theoretical conclusions often rested on little or flimsy experimental support; and what experimental support was present was often derived from the work of others.

Davis gives as an example Lavoisier's statement (in modern terms) that although all of ordinary air cannot be converted to carbon dioxide via combustion or be used to produce the calcination (or oxidation of metals), some other gas is mixed with air that can support these processes. When that other gas is exhausted, the oxidation ceases. To return to quoting Davis (1966, p. 136): "His only laboratory evidence for this was a single experiment in which the enclosed atmospheric air was diminished by a mere 5 percent when calcination of the enclosed metal (it was lead) ceased. . . . We now know that the diminution, in a properly conducted experiment, should have approximated 20 percent, the percentage of the atmosphere composed of oxygen. Priestley, in an experiment published before Lavoisier conducted his, had come much nearer the mark, noting a diminution in air volume of between one-fifth and one-fourth." Called to account for this discrepency, Lavoisier was later to claim that the amount of "fixable elastic fluid" might well vary over time and space.

In another statement in *Opuscules*, Lavoisier's informal subjectivity again outran his experimental results. Lavoisier stated that calcination in an enclosed vessel could take place only to the extent that there was air in the vessel. Further, if the vessel were exhausted of air, if calcination took place at all it would be very different from that which occurs in the presence of air and would not result in an increase of weight in the calx. But the experiments that supported these statements were carried out only after the book was published.

Similar anticipation of experimental findings occurred in the November 12, 1774 memoir about experiments with sealed vessels, which involved weighing the vessel before and after combustion. Although Lavoisier claimed to have completed many successful experiments with tin and lead, in fact many of the sealed retorts he was using exploded so that he could draw no conclusions from such trials. Indeed, Davis (1966, p. 143) tells us that "[h]e had actually completed two conclusive experiments with tin and, as he later confessed, 'hardly one for lead.'"

Although he frequently wrote that careful experimentation was crucial to good science and even more frequently characterized his measurement as "careful" and his data as "accurate," on occasion Lavoisier was also apt to play fast and loose with his experimental data in the service of his theory. For example, Lavoisier claimed in the November 12, 1774 memoir that he very carefully weighted the sealed retorts both before and after heating, finding the increase in weight after opening the retort always to be exactly equal to the amount of weight gained by the metal. But such precise equality of weights would not actually be found. A film of water vapor is always present on cool glass in atmospheric air. Hence Lavoisier should have noticed a slight decrease in weight after heating. He did notice a slight gain in weight in one retort that he left standing overnight (and which thus cooled) but attributed no significance to that gain, not being aware of the water vapor film.

Thus we see Lavoisier ignoring the details of his data—a slight weight loss immediately after heating and a subsequent slight weight gain—that were inconsistent with his contention that the weight of the closed system remained the same. He may well have subjectively decided that the weight is indeed constant and that the variations from constant weight were mere experimental artifacts. His personal beliefs were thereby injected into his experimental results. Again, the remarkable thing about these informal uses of subjectivity in scientific method is that these great scientists had such keen understanding of the processes underlying their researches that they generally got their conjectures right!

Another example of such data bending in the service of a theory cited by Davis (1966, p. 147) is worth quoting. To show that *mercurius calcinatus per se* (the red powder obtained by heating mercury by itself, without the presence of charcoal) was a true metal calx, Lavoisier, in the Easter memoir, reported heating it in a sealed retort, and in 2.5 hours reducing it to mercury and "air": ". . . 1 ounce contained 576 grains, and when the 'operation' was 'completed,' he found that the 576 grains of calx originally placed in the retort had been 'reduced' to 522 grains of liquid mercury, a net loss of 54 grains of solid matter. Simultaneously there had been 'set free' and collected above the water in the bell jar 78 cubic inches of air. If it were assumed that the 54 grains of solid calx had been transformed into 78 cubic inches of air, 'then each cubic inch [of air] must weigh a little less than two-thirds of a grain, which does not differ much from the weight of ordinary air.'"

But Conant (1956) pointed out that Lavoisier had earlier said that a cubic inch of atmospheric air weighed 36/78 of a grain. He was now calculating that a cubic inch of atmospheric air weighed 54/78 grain. Thus the agreement

Lavoisier claimed was due considerably more to the influence of his theory that expected the results to be the same than to the actual laboratory results. His expectation was that this new air was a particularly pure sample of atmospheric air.

In yet another instance of the same sort, Guerlac (1973) describes an experiment carried out on fermentation (for additional discussion of *fermentation*, see Pasteur; Section 4.10). Lavoisier assumed that the amounts of carbon, hydrogen, and oxygen in the original sugar he used would be equal to the amounts of these elements contained in the products of the reaction, i.e., alcohol, carbon dioxide, and acetic acid. Although his data were unreliable, his end result somehow was correct. Guerlac (1973, p. 83) quotes Arthur Harden as noting in his *Alcoholic Fermentation*: "The research must be regarded as one of those remarkable instances in which the genius of the investigator triumphs over experimental deficiencies." And because the data reveal grave errors, "it was only by a fortunate compensation of these that a result near the truth was attained."

Kenneth Davis, whose 1966 work has supplied us with much of the material attesting to Lavoisier's use of subjectivity, eventually exonerates him of real wrongdoing. Lavoisier had a kind of intellectual pride that kept him from fully acknowledging intellectual debts but also kept him from fully ignoring them. It also kept him stubbornly trying to find the laboratory results he expected. Davis writes (1966, p. 158): "He could not resist temptations to fake experimental evidence just a little bit now and then, in his eagerness to publish conclusions he knew must be true since they were logically implied by other firmly established truths; nor could he resist the felt need, rooted in his pride, to correct such falsifications. He redid his experiments until they were successful, and when they at last justified the conclusions he had prematurely published or when they required some modification of these, he revised and corrected his memoirs."

REFERENCES

Conant, James Bryant (1956). *The Overthrow of the Phlogiston Theory*, Vol. 2 of Harvard Case Studies in Experimental Science. Cambridge, MA: Harvard University Press.

Davis, Kenneth S. (1966). *The Cautionary Scientists: Priestley, Lavoisier, and the Founding of Modern Chemistry*. New York: G.P. Putnam's Sons.

Donovan, Arthur (1993). *Antoine Lavoisier*. Cambridge, MA: Blackwell.

Guerlac, Henry (1961). *Lavoisier—The Crucial Year: The Background and Origin of His First Experiments on Combustion in 1772*. Ithaca, NY: Cornell University Press.

——— (1973). "Lavoisier," in Charles Coulston Gillispie (Ed.), *Dictionary of Scientific Biography*. New York: Chanles Scribner's Sons.

Holmes, Frederic Lawrence (1985). *Lavoisier and the Chemistry of Life: An Exploration of Scientific Creativity*. Madison, WI: University of Wisconsin Press.

————(1987). "Scientific Writing and Scientific Discovery," *Isis*, **78**:220–235.

————(1988). "Lavoisier's Conceptual Passage," *Osiris*, **4**:82–92.

————(1990). "Laboratory Notebooks: Can the Daily Record Illuminate the Broader Picture?" *Proceedings of the American Philosophical Society*, **134**:349–366.

McKenzie, A. E. E. (1960). *The Major Achievements of Science: The Development of Science from Ancient Times to the Present*, New York: Simon & Schuster.

McKie, Douglas (1952). *Antoine Lavoisier: Scientist, Economist, Social Reformer*. London: Constable and Co.; New York: Henry Schuman.

Susac, Andrew (1970). *The Clock, the Balance, and the Guillotine: The Life of Antoine Lavoisier*. Garden City, NY: Doubleday.

4.7 ALEXANDER VON HUMBOLDT (1769–1859)

Alexander von Humboldt was a naturalist—a climatologist, geographer, geologist, and geomagnetist—as well as a botanist, ecologist, and explorer.

4.7.A Brief Biographical Sketch

To sketch Alexander von Humboldt's career is to chronicle almost 60 years of scientific activity. A son of nobility, von Humboldt was born in Berlin in 1769. With his brother, Wilhelm who became a famous philologist, he was educated by private tutors before going on to the university. Besides the classics and modern languages, his early training included botany. After a brief time at the University of Frankfurt an der Oder, both von Humboldt brothers moved to the University of Göttingen. Although at this point self-trained in geology and mineralogy, Alexander undertook what he called a two-month-long "scientific tour" throughout Germany, studying botany and geology, which resulted in his first scientific paper, "Mineralogical Observations on Some Basalts of the Rhine."

    This early Rhine expedition was to be only the first of many such scientific journeys. In Göttingen, Alexander met George Adam Forster, who had accompanied his father on Captain Cook's second voyage around the world and written eloquently and excitingly about the voyage; surely, Forster whetted von Humboldt's appetite for scientific travel. In fact, the two young men embarked on a several-month-long trip on the lower Rhine and to France and England.

At his mother's insistence, von Humboldt next enrolled in the Hamburg School of Commerce, but from there wrote to A. G. Werner asking admittance to the Freiberg Mining Academy, which was then an important scientific institution under Werner's directorship. Although noting that he had only six months to devote to studies there before having to enter employment, von Humboldt was indeed admitted. (Indeed, the prospective student asked for and received assurances that following his course of study at Freiberg he would be employed as an inspector by the Department for the Regulation of the Mining and Smelting Works.) At Freiberg, von Humboldt followed a curriculum dealing with both the practical and scientific aspects of mining while continuing his botanical investigations with a concentration on subterranean plants, reading in physics, and embarking on a self-directed study of chemistry.

Following his time at Freiberg, in 1792, von Humboldt did receive an appointment as assessor in the Administrative Department of Mines and Smelting Works, in which position he made tours of inspection. After only six months he was promoted to Superintendent of Mines in the Franconian Principalities. His official duties again involved travel, and these travels were occasions for making observations and collecting data. He was readying for publication a manuscript on underground plants and investigating salt springs. While he was superintendent he established a free school for the miners to educate them about dangers in the mines and on how to recognize strains of ores; he long hoped, however, to leave public service so that he could travel more extensively. When his mother died in 1796, he was at last free to do so. He resigned his position and began to plan his travels in earnest.

He spent some time in Vienna and in Paris, preparing himself for his travels by studying botany and learning skills of geodetic, meteorological, and geomagnetic measurement. During the 1790s, von Humboldt became acquainted with Goethe and Schiller. Both a plan to sail around the world and a projected expedition to Egypt were frustrated by the political upheavals of the Napoleonic Wars. Finally, he received permission to travel to the Spanish colonies in the New World, and set sail in 1799. He was accompanied on what turned out to be a five-year journey by Aimé Bonpland, a French botanist whom von Humboldt had met in Paris.

During the sea journey to the New World, von Humboldt and Bonpland spent their time making astronomical and meteorological observations, measuring the temperature and chemical composition of the ocean, and examining marine life that was brought on board. Landing first in the Canary Islands, the two explorers climbed the volcanic peak of Teide and examined and collected botanical and geological specimens. From there the party sailed to

Cumaná, Venezuela, where von Humboldt and Bonpland gathered more botanical specimens, made more meteorological and astronomical observations, and tested their instruments. They made a trip to an inland mountain mission at Caripe, exploring caves, finding a new species of bird, and collecting and cataloging botanical specimens. After experiencing their first earthquake and meteor shower, the travelers set off for Caracas, where they rented a house, and while waiting out the rainy season, ascended Silla, the highest of the peaks surrounding Caracas. Finally, they headed for the interior, the torrid plains of the llanos (where von Humboldt recorded a ground temperature of 50°C at their noon arrival). Visiting a cattle-trading station, the travelers experimented with electric eels; and finally, arrived at the Rio Apure, a major tributary of the Orinoco. Traveling in a canoe with Indian paddlers and a pilot, they proceeded down the Rio Apure and up the Orinoco, portaging past the great cataracts where no men of science had yet penetrated. The jungle was filled with wildlife, including abundant and fierce jaguars and incessantly annoying insects. Bonpland collected plants, von Humboldt made constant measurements, and they collected a small menagerie of animal specimens. The purpose of the journey was primarily to establish the existence of the Casiquiare Canal, a natural waterway that joins the Rio Negro with the Orinoco, thus connecting two huge river systems (the Orinoco and the Amazon). This accomplished, they headed downstream on the Orinoco, stopping on their way down the coast to investigate the composition and effects of curare.

At last they arrived at Angoustura in the river straits and traveled overland back to Cumaná; Botting (1973, pp. 36–37) summarizes their journey thus: ". . . the first scientific exploration of 1500 miles of almost unknown territory between the headwater areas of the Amazon and Orinoco basins. They had measured the latitude and longitude of more than 50 places, including the Casiquiare canal, taken a series of important magnetic readings, and made a huge collection of plants, 12,000 specimens in all, many of them rare or new to science."

The next stage of their South American journey took the travelers to Cuba and thence to Cartagena and overland to Bogotá and Quito. Staying in Quito for six months, von Humboldt and Bonpland investigated the surrounding volcanoes, in the process learning about and recording the symptoms of altitude sickness. They climbed Chimborazo, making seismographic measurements and mapping the geography of plants, and at 20,702 feet set an altitude record that remained unbroken for 30 years. Thence they traveled to Lima and sailed to Guayaquil, making measurements of the cold current that has since been named after von Humboldt. From Guayaquil the party sailed to Acapulco and spent almost a year traveling around Mexico. This period was filled less with field research and more with academic research on the geography, economics, and politics of the country and eventually resulted in the first modern volume of regional geography, von Humboldt's *Political Essay on the Kingdom of New Spain.*

The next port of call was Havana, and thence to the United States to visit Philadelphia, the scientific capital of the nation and Washington, where von Humboldt visited with President Thomas Jefferson before sailing back to France. With the exception of trips such as to Berlin and to Italy to observe the eruption of Vesuvius, von Humboldt chose to live in Paris for the next quarter century. Here he arranged the enormous collections he had brought back from the New World, wrote up his results in what became 30 volumes on the South American trip, lectured, did experimental work with Louis Gay-Lussac on the composition of the atmosphere, and planned a future expedition to Asia. He enjoyed a very active social life, meeting all the eminent scientists of the day, and being much in demand by the rich and famous. During these years he tried unsuccessfully to mount a journey to Asia, to examine the Himalayas, but was refused entry permits by the British. He expended all his personal funds on the publication of his books; to support him the Prussian king, Frederick William III, appointed him court chamberlain, a post that was a sinecure but paid a stipend. Von Humboldt traveled to England and throughout Italy as an advisor to the king. In 1827, at the insistence of Frederick William, von Humboldt returned from Paris to Berlin to take up the post of court chamberlain in earnest, a post he filled as a sort of cultural advisor, at the beck and call of the king.

In 1829 von Humboldt was finally able, at the invitation of Czar Nicholas, to make at least part of the long-planned journey of exploration to Asia. He and two companions (Christian Ehrenberg and Gustav Rose) in nine months traveled 10,000 miles by carriage across Russia and to the Altai Mountains at the border of China. Thence they explored the Urals and the great steppe of Ishim and returned to the Caspian Sea. Unlike the earlier expedition to South America, this journey was carried out in comparative comfort and with many official receptions, but throughout, von Humboldt made geographical, geological, and meteorological observations. These were published primarily as *Asie Centrale, recherches sur les chaînes des montagnes et la climatologie compaerée* (Central Asia: Research on Mountain Chains and Comparative Climatology, 1843).

The last 30 years of von Humboldt's life were spent primarily in Berlin (with yearly visits to Paris to renew acquaintance with prominent scientists), writing up the results of his investigations and working on his magnum opus, *Kosmos*. During this time he was also instrumental in convincing the British government to set up a series of observatories in Canada, South Africa, Australia, and New Zealand and to send an expedition to Antarctica to measure the earth's magnetic field. The data gathered at these observatories, together with data already being gathered from a chain of observatories organized by Carl Friedrich Gauss in Germany, England, and Sweden, were for the purpose of determining whether the magnetic storms von Humboldt had observed were of terrestrial or extraterrestrial origin. It was from these data that Sir Edward Sabine later concluded that the appearance of magnetic storms in Earth's atmosphere was correlated with the changing activity of sunspots. Von Humboldt died in Berlin on May 6, 1859, still at work on the fifth volume of *Kosmos*.

4.7.B   Von Humboldt's Scientific Contributions

Following de Terra (1955, p. 374) we give a list of von Humboldt's major contributions (with the fields of application in alphabetical order):

1. *Anthropology*. Von Humboldt made ethnologic observations on American aborigines and traced, through such artifacts as calendars and myths, their origins to northeast Asia. Von Humboldt's geographic orientation suggested that environmental factors influenced native customs, as they influenced botanical and zoological features.

2. *Astronomy*. Von Humboldt made careful observations on the meteor showers of 1799 and helped to establish the periodicity of such showers. The first volume of *Kosmos* was largely devoted to a description of the planets and cosmic nebulae.

3. *Botany*. Von Humboldt collected and described enormous numbers of plant specimens, many of them new species. His theory of botanical zones, both across Earth's latitudes and across the elevations of mountains, constituted the beginning of plant geography.

4. *Geography*. Von Humboldt was the founder of modern geography and the originator of the use (and naming) of isotherms.

5. *Geology*. Von Humboldt recognized the relationship between volcanism and earth structures and noted the similarities between geologic formations across the continents he explored. His climb of Chimborazo established an altitude record that stood for many years.

6. *Geophysics*. Von Humboldt discovered the law of declining magnetic intensity between the earth's poles.

7. *Physiology*. Von Humboldt was the first to observe the effects of high altitude on the human body; research growing from these careful observations contributed to the development of manned flight.

8. *Zoology*. Von Humboldt's travels in South America led to the description of such animals new to science as the electric eel, the alligator, and the monkey. He pioneered the practical use of guano.

Von Humboldt also made important contributions to meteorology and oceanography—in fact, the Humboldt Current is named after him.

Meadows (1987, p. 109) describes von Humboldt's contributions:

The scientific revolution of the 17th century was much concerned with attempts to interpret the world about us in terms of mechanical models. During the 18th century this approach proved increasingly useful in "natural philosophy"—such sciences as physics and chemistry, but it was of much less value for studying "natural history"—

environmental or biological problems. The scientific revolution that culminated in Newton therefore gave greater aid to natural philosophy than to natural history. Humboldt's importance is that he helped focus attention on the significance of scientific explanation in natural history. He linked the world view of Newton in the 17th century with that of Darwin in the 19th century.

There is no question that von Humboldt was regarded as eminent in his own time. He is often referred to in his time as second in fame only to Napoleon. Von Hagen (1945, p. 168) quotes the American writer Bayard Taylor (1825–1878) writing in 1857 to say: "I came to Berlin not to visit its museums and galleries, its operas, its theatres, nor to mingle in the gay life—but for the sake of seeing and speaking with the world's greatest living man—Alexander von Humboldt...." On the same page, von Hagen refers to Charles Darwin as von Humboldt's spiritual protégé and quotes Darwin as calling von Humboldt "the greatest scientific traveler who ever lived." Thus von Humboldt derives his eminence from his scientific travels, collections, and descriptions. An enormous scientific curiosity and the impetus to classify and understand drove von Humboldt.

4.7.C Major Works

Survey Measurements
von Humboldt, Alexander (1808). *Conspectus Longitudinum et Latitudinum Geographicaraum.* Paris: F. Schoell.

——(1808–1810). *Recueil d'Observations Astronomiques, d'Opérations Trigonométriques et de Mesures Barométriques. Rédigées et Calculées par Jabbo Oltmanns,* 2 vols. Paris: F. Schoell and Tubingen: J. G. Cotta.

——(1809). *Nivellement Barométrique.* Paris: F. Schoell.

Botany
C. S. Kunth (1822–1826). *Synopsis Plantarum,* 4 vols. Paris: Apud F. G. Levrault.

von Humboldt, Alexander (1805–1817). *Plantes Équinoxiales,* 2 large folio vols., 144 plates. Paris: F. Schoell and Tubingen: J. G. Cotta.

——(1806–1823). *Monographie des Melastomacées,* arranged by A. Bonpland; 2 large folio vols., 120 color plates. Paris: Gide.

——(1816–1825). *Nova Genera et Species Plantarum,* described by Carl Sigismund Kunth; 7 large folio vols., 700 plates, mostly in color. Paris: Sumtibus Librariae graeco-latino-germanicae.

——(1819–1824). *Mimoses et Autres Plantes Légumineuses,* described by C. S. Knuth; 2 large folio vols., 60 color plates. Paris: Librairie greque-latine-allemande.

——(1829–1834). *Révision des Graminées,* 2 large folio vols., 220 color plates. Paris: Gide fils.

Plant Geography
von Humboldt, Alexander (1805). *Essai sur la Géographie des Plantes.* Paris: F. Schoell.

Zoology and Comparative Anatomy

von Humboldt, Alexander (1805–1833). *Recueil d'Observations de Zoologie et d'Anatomie Comparée*, 2 vols., 57 plates. Paris. Vol. 1 published Paris 1805 Levrault, Schoell; Vol. 2 published 1833 Paris: C. J. Smith, Gide.

Travel and Geography

von Humboldt, Alexander (1808–1812). *Atlas Géographique et Physique du Royaume de la Nouvelle-Espagne*, large folio, 20 maps. Paris: F. Schoell and Tubingen: J. G. Cotta.

———(1813). *Altlas Pittoresque—Vues des Cordillères et Monuments des Peuples Indigénes de l'Amérique*, 2 large folio vols., 69 color and monochrome plates. Paris: F. Schoell.

———(1811). *Essai Politique sur le Royaume de la Nouvelle-Espagne*, 2 vols. Paris: F. Schoell.

———(1814–1825). *Relation Historique du Voyage aux Régions Équinoxiales du Nouveau Continent [in which is included the political essay on the island of Cuba]*, 3 vols. Paris: Schoell. This was published in England as *Personal Narrative of Travels to Equinoctial Regions of America*, translated by Thomasina Ross, 3 vols., London, 1852.

———(1814–1834). *Atlas Géographique et Physique des Régions Équinoxiales du Nouveau Continent*, large folio, 40 plates. Paris: F. Schoell: Gide.

———(1814–1834). *Examen Critique de l'Histoire de la Géographie du Nouveau Continent*. Paris: Gide.

Other Major Works

von Humboldt, Alexander (1843). *Asie Centrale, Recherches sur les Chaînes des Montagnes et la Climatologie Comparée*, 3 vols. Paris: Gide.

———(1845–1862). *Kosmos*, 5 vols. Stuttgart: Cotta. Translated by E. C. Otté (1852) as *Cosmos: A Sketch of a Physical Description of the Universe*. New York: Harer.

4.7.D   Subjectivity in the Work of von Humboldt

As we have seen, Alexander von Humboldt's methodological approach to science was largely to observe carefully, to record his observations, and often, to develop a taxonomy about the entities in the phenomenon he was observing. Such a methodological approach was, of course, entirely appropriate for a naturalist. As a consequence, however, he does not appear to have often proceeded beyond such initial stages to the later stages of the scientific method proposed by Galileo: hypothesizing, experimentation, theory formulation, more experimentation, revision of theory, prediction, and so on. For this reason, many of his conclusions about the world resulted largely from conjecture, subjective speculation, and an Aristotelian view of the purposefulness, and harmony and unity in the universe. We will see this, for example, in von Humboldt's position on Neptunism, discussed below.

At the time von Humboldt was trained, and trained himself, in geology and mineralogy, the dominant theory of the formation of rocks was termed *Nep-*

tunist, so named because it traced that origin to the actions of waters in oceans and rivers, with rocks being precipitated out. (This was in opposition to the *Plutonist school*, which traced the origin of Earth's rocks to the actions of volcanoes and other movements of the earth.) The *Neptunist school* was also referred to as the *Wernerian school* after Abraham Gottlob Werner, a very prominent geologist of his day and director of the Freiberg Mining Academy, where von Humboldt studied. Not only was the Neptunist theory widely accepted in scientific circles, it was also congruent with religious beliefs. Plutonism was reviled because it preached continuing violent changes in Earth's crust, unlike Neptunism, which saw a single creation and then only gradual wearing away with water—the Neptunist view was much more consistent with religious beliefs.

Von Humboldt's first paper, "Mineralogical Observations on Some Basalts of the Rhine," written some years before he attended the Freiberg Mining Academy under Werner, was premised on the Neptunist theory of the origin of the earth's rocks. Indeed, von Humboldt wrote a letter dated July 25, 1790 to Werner at Freiberg (quoted in Bruhns, 1873, p. 94) that said in part:

> In my journey to England, whence I have just returned, I traversed for a second time the mountains of the Rhine. I found nothing to necessitate the supposition of previous volcanic agency, but on the contrary abundant evidence to prove the aqueous formation of basalt. Your theory that a stratum of basalt had once covered the whole earth never seemed to me more reasonable or more obvious than at Linz and Unkel, where I noticed horizontal layers of basalt upon the highest summits. I shall be severely censured for this confession by many of our geologists, and my pamphlet (if it not be altogether overlooked) will be subjected to no tender criticism. But such considerations shall never prevent me from saying what I feel to be true, and I trust that I shall always maintain this resolution. I have resided for a considerable time in a region that has been called volcanic, and I have industriously traversed on foot the mountains of Hanover, Hesse, and the Rhine, as well as those in the neighborhood of Zweibruken; but I cannot accept the hypothesis so charmingly set forth by De Luc in his geological work "Lettres physiques et morales."

This is an instance of a strong belief and intuitive informed understanding held prior to observing data—that basalts are formed by sedimentary action rather than by volcanoes or other upheavals of the earth—influencing von Humboldt to see that which he expected to see.

Werner's Neptunists ideas were later discredited in the face of the geologic theory proposed by James Hutton as early as 1785, which maintained that basalt and granite were volcanic and magmatic, as is recognized by modern geologists. That theory was later expanded and popularized by Sir Charles Lyell. Von Humboldt himself abandoned his strong prior belief in the face of overwhelming evidence. He became convinced of the correctness of the Plutonist stance during his time exploring the volcanoes around Quito, when he saw the difference between eruptive and sedimentary rocks and examined

the change that can take place in the structure of minerals in the neighbor-
hood of a volcanic explosion. That his original prior belief affected his observations was clear to von Humboldt himself. Lowenberg in Bruhns (1872, vol.
1, p. 306) points out that von Humboldt's first publication of his observations
during the ascent of volcanoes were published under such separate rubrics as
botany and geography, and only later did he compose separate topographies
of the ascents. The reason for waiting to present this kind of integrated treatment was "that he thought it necessary to wait until he could show 'the relative worthlessness' of his labours, by distinguishing between those geological
observations, which had been made on principles since proved obsolete, and
those of a character which could not be affected by time."

Von Humboldt's change of views was consequential for him. For example,
a friendship between von Humboldt and Johann Wolfgang von Goethe began
in the 1790s and continued until Goethe's death in 1832. Von Humboldt
absorbed a good deal of Goethe's world view and admired his scientific
research as well as his literary achievements. But with this friendship he was
flirting with a very critical force. The *Encyclopaedia Britannica* (1975, Vol. 8,
p. 229) tells us of Goethe that: "Few have been as aware of the mental
processes involved in the study of natural phenomena; few have been more
alive to the hazards that beset the scientist, at every level, from sheer observation to the construction of a theory; and few have been more conscious of
the unwitting theorizing involved in even the simplest act of perception."

Goethe admired von Humboldt's observations, but he roundly criticized
his friend when von Humboldt abandoned his adherence to the Neptunist
theory of the formation of rocks. In fact, two verses in Goethe's *Faust* speak
to the point, and were widely seen at the time as scientific criticism of von
Humboldt. In the first of these verses, Faust praises the Neptunist view, while
in the second verse, Mephistopheles himself ridicules the Plutonists. Goethe
acknowledged, in our terms, the hold a strong, if mistaken, prior belief had
upon him: "I have spent several years of my life in studying the mutual
connection of the stratified formations. The views I formed coincided with
Werner's doctrine, and I continued to hold them even after I had good reason
to believe that they left many problems unresolved" (quoted in Bruhns, 1873,
vol. 1, p. 168).

Also of concern to von Humboldt were the sensibilities and reputation of
his mentor, A. G. Werner. At the 1850 commemoration of the centenary of
Werner's birth, von Humboldt credited the director of the Freiberg Mining
Academy with a strong influence on his thinking and career and commented
that he (von Humboldt) had endeavored to "elevate his works, in modern
times so often misunderstood, to their right position" (Bruhns, 1973, vol. 1,
p. 115).

More broadly, early in his scientific career, von Humboldt resolved to
uncover a harmony in nature that he believed to be omnipresent, with all
natural objects being interrelated. He articulated that view and his aim to
confirm it as he set sail for South America in 1799: "In a few hours we sail

round Cape Finisterre. I shall collect plants and fossils and make astronomic observations. But that's not the main purpose of my expedition. I shall try to find out how the forces of nature interact upon one another and how the geographic environment influences plant and animal life. In other words, I must find out about the unity of nature" (quoted in Botting, 1973, p. 64). In 1844, in the preface to *Kosmos*, von Humboldt reiterates:

> Although the outward relations of life, and an irresistible impulse toward knowledge of various kinds, have led me to occupy myself for many years—and apparently exclusively—with separate branches of science, as, for instance, with descriptive botany, geognosy [a branch of geology], chemistry, astronomical determinations of position, and terrestrial magnetism, in order that I might the better prepare myself for the extensive travels in which I was desirous of engaging, the actual object of my studies has nevertheless been of a higher character. The principal impulse by which I was directed was the earnest endeavor to comprehend the phenomena of physical objects in their general connection, and to represent nature as one great whole, moved and animated by internal forces.

A quote from Van Dusen (1971) offers insight into the early training of Alexander von Humboldt and the impact of that training on his later achievements and his effort to demonstrate the unity of nature. On page 18 we read: "[I]n each of young Humboldt's major interests one can discern a definitely similar pattern. It is one of unity that transcends diversity; this unity reconciles various forms so that they appear to us aesthetically harmonious and coherent. . . . In order to realize the fundamental unity that encompasses the universe, one must be able to study one's environment and reflect upon it philosophically."

Nelken (1980, p. 87) shows us how this vision of unity extended to von Humbold's view of human beings and their culture:

> [H]e was the first to make on-the-spot sketches of the monuments of the indigenous American people and to point out the significance of their vanished culture and civilization. Although Humboldt was brought up on Winckelmann's theory of the artistic and intellectual superiority of ancient Greece and Rome, he did not disregard the relics of the Incas and the Aztecs, which were so different from the classic canons of beauty. These "savage" images, with their "primitive" proportions, formed an integral part of their natural environment, thus confirming Humboldt's theory of the unity of nature, peoples, and their creations. He considered these monuments an important source, along with those of Egypt, Greece and the Orient, for comparative studies of the cultural development of mankind.

Bruhns tells us that this world view was part of the basis of the friendship between von Humboldt and Goethe. We may speculate on the effect of von Humboldt's conviction of the unity and harmoniousness of nature. Such a deeply held and broadly conceived view must surely have influenced not only von Humboldt's collection, selection, and organization of his own data, but also his choice of which data of others to include (in such works as *Kosmos*) and how to treat them. Harmony he expected, and harmony he found.

REFERENCES

Botting, Douglas (1973). *Humboldt and the Cosmos.* New York: Harper & Row.

Bruhns, Karl (Ed.) (1873). *Life of Alexander von Humboldt Compiled in Commemoration of the Centenary of His Birth* by J. Lowenberg, Robert Avé-Lallemant, and Alfred Dove, 2 vols., translated by Jane and Caroline Lassell. London: Longmans, Green.

de Terra, Helmut (1955). *Humboldt: The Life and Times of Alexander von Humboldt, 1769–1859.* New York: Alfred A. Knopf.

Encyclopaedia Britannica (1975). Vol. 8.

Klencke, H. (1852). *Alexander von Humboldt: A Biographical Monument,* translated by Juliette Bauer. London: Ingram, Cooke.

Meadows, Jack (1987). *The Great Scientists.* New York: Oxford University Press.

Nelken, Halina (1980). *Alexander von Humboldt: His Portraits and Their Artists: A Documentary Iconography.* Berlin: D. Reimer.

Stearn, William T. (Ed.) (1968). *Humboldt, Bonpland, Kunth and Tropical American Botany: A Miscellany on the "Nova Genera et Species Plantarum."* Stuttgart, Germany: Verlag von J. Carmer.

Van Dusen, Robert (1971). *The Literary Ambitions and Achievements of Alexander von Humboldt.* Berne, Switzerland: Herbert Lang; Frankfurt am Main, Germany: Peter Lang.

von Hagen, Victor Wolfgang (1945). *South America Called Them: Explorations of the Great Naturalists: La Condamine, Humboldt, Darwin, Spruce.* New York: Alfred A. Knopf.

4.8 MICHAEL FARADAY (1791–1867)

Michael Faraday was a natural philosopher, physicist, and chemist who is known principally for showing that an electric current could be induced to flow in a coil of wire by moving a magnet through the coil. He invented the electric motor and the dynamo, was one of the originators of field theory, and was among the founders of electrochemistry.

4.8.A Brief Biographical Sketch

Michael Faraday was born in Surrey, England, to a father who was a member of the Sandemanian church, and a mother who although not a member (i.e., she had not made a confession of faith), belonged to the congregation. His father was a blacksmith and the family lived in the humblest of circumstances. Indeed, at the age of 9, Michael had to make do weekly with a single loaf of bread from his parents. He attended a common day school and had what he called a "most ordinary" education, consisting of only a single year of formal schooling.

His father having died, and facing the need to help support his widowed mother and younger sister, beginning in 1804, Faraday was an errand boy and then an apprentice to George Riebau, who trained him as a bookbinder. In the bindery, young Faraday educated himself by reading the scientific books that were being bound, as well as books from Riebau's personal library. In particular, in his writings, Faraday mentions Marcet's *Conversations in Chemistry* and the electrical treatises in the *Encyclopaedia Britannica*. During this time, Faraday also conducted simple experiments in chemistry, constructed pieces of electrical apparatus in his attic room in the home of Riebau, and also attended public lectures on natural philosophy. Perhaps most important, in 1812 Faraday heard four lectures on chemistry, electricity, and electrolysis by Sir Humphry Davy at the Royal Institution and wrote up and bound copious notes about what was taught.

After Faraday became a journeyman bookbinder, he sent those notes to Davy, expressing his wish to "enter into the service of science" (Jones, 1870, vol. 1, p. 47) and asking for Davy's help in making that entrance. Within months, Davy had Faraday appointed his assistant at the Royal Institution; he joined the Royal Institution at the age of 22 in 1813 and was to remain there for his entire scientific career. Indeed, his living quarters were thenceforth in the Institution's building. Six months later, Faraday accompanied Davy on a year-and-a-half-long trip to the continent as his assistant and secretary.

Working under Davy, Faraday accomplished the liquefaction of several gases and discovered two new chlorides of carbon. In the momentous year of 1821, he first made a magnetic needle revolve around an electric current (later he was to induce an electric current by inserting and removing a magnet in a coil of wire) and married Sarah Barnard. Although the marriage was childless, it seems to have been extremely happy.

In 1824, Faraday was elected a Fellow of the Royal Society, in 1825 he was appointed director of the Royal Academy, and in 1833 he became Fullerian professor of chemistry there. He used his position to give many very vivid public lectures on science; his special lectures for children were especially well received, and the institution of those lecture continues today. The 1820s and 1830s were a period of extreme productivity, so much so, in fact, that Faraday suffered from exhaustion, and on doctor's orders in 1840 and 1841 he cut back on his scientific endeavors. He returned to his research thereafter and continued for another quarter century to make momentous discoveries. He resigned as superintendent of the Royal Institution in 1866 and died the following year, just before his seventy-sixth birthday.

Perhaps the most salient nonscientific fact about Faraday is his adherence to Sandemanianism, a sect with which his grandfather, father, and wife were also involved. He made a confession of faith in 1821 and was elected an elder in 1840. The sect followed extremely demanding religious practices. Its adherents pledged to live purely and according to the Bible, to avoid the amassing of worldly goods, and to refrain from games of chance, for lotteries were regarded as sacred means of division of resources, as often mentioned in scripture. Sandemanians believed that the laws of God that governed everyday life were all available in the Bible, so that Bible reading and commentary were necessary to the adherents in order to find the necessary guidance. There was no clergy to intervene or make exegesis.

The religion stressed the intelligibility, beauty, and symmetry of the universe, and hence has been seen (e.g., by Williams, 1965) as a root of Faraday's science. Indeed, Faraday himself wrote: "[E]ven in earthly matters I believe that the invisible things of HIM from the creation of the world are clearly seen, being understood by the things that are made, even His eternal power and Godhead . . ." (cited by Cantor, 1985, p. 71). The notions of simplicity in God's creation, God's avoidance of unnecessary complexity, and man's ability to read the book of God were paralleled, for Faraday, by an ability to read the book of nature, also written by God (Cantor, 1985). It was man's moral duty to learn about the world, and although God would shun unnecessary complexity, He would not necessarily make the learning easy for man.

4.8.B Faraday's Scientific Contributions

Originally working in chemistry, Faraday moved through an interest in electrochemistry and the development of the basic laws of electrolysis to the study of electricity and magnetism. All this was in addition to public lectures and other obligations, including industrial consulting, entailed by his connection with the Royal Institution. As an analytic chemist, Faraday discovered two new compounds of carbon (tetrachloroethene and hexachloroethane). He studied the liquefaction of gases (achieving the liquefaction of chlorine, ammonia, carbon dioxide, sulfur dioxide, nitrous oxide, hydrogen chloride, and hydrogen sulfide, among others), thus demonstrating the continuity of matter, showing that the three states are convertible. He discovered what he called *bicarburet of hydrogen*, which was later named *benzene*, and isobutene, and he established the chemical formula for naphthalene. All these new molecules find modern uses in the aniline dye, chemical, explosive, and pharmaceutical industries. In 1827 Faraday published a major monograph entitled *Chemical Manipulation*.

Faraday's most important accomplishment was the discovery of electromagnetic induction, but the path leading to that discovery was long. He learned of the discovery in 1820 by Hans Oersted that a wire carrying an electric current deflects a magnetized compass needle and the further discovery by André Ampère that wires carrying electric currents exert forces on each other. Thereafter, Faraday carried out a major survey of the state of current knowledge about the relations between electricity and magnetism. He then carried Oersted's work further to show that a wire carrying an electric current could be made to rotate around a stationary magnet and a magnet could be made to rotate around a stationary wire. This first demonstration of electromagnetic rotation, converting electrical energy into mechanical energy, formed the basis for the electric motor.

But Faraday believed that further conversion of energy was also possible, that magnetism could produce electricity. He began a series of experiments that culminated in 1831 with the demonstration of electrical inductance and the enunciation of his laws of electromagnetic induction, that an electric current is set up in a closed circuit by a changing magnetic field. From this demonstration grew the development of both the transformer and the dynamo. To explain the phenomenon of inductance, Faraday proposed the concept of a field of force surrounding a magnet or a wire carrying an electric current. The physical visualization of such a field is exemplified by lines of force which iron filings trace out on a sheet of paper placed over a bar magnet. These ideas were unpopular at the time, but were later mathematized by James Clerk Maxwell and became the basis of modern field theory.

Faraday's investigations of electrolysis led to his discovery of the famous laws named for him. Thomas (1991, p. 48) quotes Faraday's own words in stating these laws:

> The first Law . . . states "Chemical action or decomposing power is exactly proportional to the quantity of electricity which passes." The second Law . . . asserts "Electrochemical equivalents coincide and are the same with ordinary chemical equivalents." In other words, the electrochemical equivalent of an element is proportional to its ordinary chemical equivalent. Today we call the amount of electricity necessary to liberate one equivalent (i.e. 1.008 grams of hydrogen and 35.457 of chlorine) from a solution of hydrogen chloride, or, in short, 1g equivalent of any element from conduction solutions or conducting molten salts of its compounds, one "faraday". . . .

Thus Faraday showed that chemical and electrical forces are intimately and quantitatively related. Other work by Faraday about the same time showed that there was no difference between the various forms of electricity then known—induced, voltaic, electrostatic, animal (as shown by the electric eel), and thermoelectricity.

Still in the 1830s, Faraday investigated the effects of different insulators on the propagation of electric currents. He discovered that each insulator had what he called a *specific inductive capacity*, a phenomenon now referred to as the insulator's dielectric constant. He believed that a substance's inductive capacity was connected to the structure of its molecules, a deduction later shown to be correct but which Faraday himself was unable to prove. In 1845, Faraday showed yet another example of the unity of forces, this time the relation between magnetism and light. Using special heavy optical glass that he had prepared decades earlier, he found that the plane of polarized light was rotated by magnetism, a phenomenon that is now called the *Faraday effect*. In 1849, Faraday tried to link the forces of electricity and gravity, but failed in the attempt.

4.8.C Major Works

Faraday, Michael (1820–1862). *Faraday's Diary. Being the various philosophical notes of experimental investigation made by Michael Faraday, DCL, FRS, during the years*

1820–1862 and bequeathed by him to the Royal Institution of Great Britain. Now, by order of the Managers, printed and published for the first time, under the editorial supervision of Thomas Martin, 7 vols. and index. London: G. Bell and Sons, 1932–1936.

———(1821–1857). *Experimental Researches in Chemistry and Physics*. Reprinted from the Philosophical Transactions of 1821–1857, *Journal of the Royal Institution; Philosophical Magazine*, and other publications. London: R. Taylor and W. Francis, 1859.

———(1831–1852). *Experimental Researches in Electricity*. Reprinted from the *Philosophical Transactions* of 1831–1852 with other electrical papers from the *Quarterly Journal of Science, Philosophical Magazine, Proceedings of the Royal Institution*, 3 vols. London: R. Taylor and W. Francis, 1839–1855.

4.8.D Subjectivity in the Work of Faraday

Faraday's religion of Sandemanianism seems to have been a major source of the informal subjectivity in his scientific work. The integrated view of the world as God's creation explicit in that religion seems to have led Faraday to seek for integration in his science as well, and in particular to believe in the simplicity of the integration of forces. Thus, having seen evidence of the magnetic field surrounding a wire conducting electricity, Faraday believed for a decade that magnetism could produce electricity. He maintained this belief in the face of numerous unsuccessful experiments, and finally was able to demonstrate the phenomenon only when his other research led him to notice the transient electric currents that occur when magnetism is turned on or off.

Faraday went further in his speculations about the identity of forces and the simplicity of creation. As noted above, he formulated the equivalence of chemical and electrical forces and proved the identity of all forms of electricity. As early as 1816, he wrote: "[T]hree apparently distinct kinds of attraction; the attraction of gravitation, electrical attraction and magnetic attraction . . . appear . . . to be sufficient to account for all the phenomena of spontaneous approach and adherence with which we are acquainted. . . . The Science of Chemistry is founded upon the cohesion of matter and the affinities of bodies and every case either of cohesion or of affinity is also a case of attraction. . . . That the attraction of aggregation and chemical affinity are actually the same as the attraction of gravitation and electrical attraction I will not positively affirm but I believe they are" (Faraday, 1816a, pp. 30–31).

Years later, in 1845, he wrote: "I have long held an opinion, almost amounting to conviction, in common I believe with many other lovers of natural knowledge, that the various forms under which the forces of matter are made manifest have one common origin; or, in other words, are so directly related and mutually dependent that they are convertible, as it were, one into another, and possess equivalents of power in their action. In modern times the proofs of their convertibility have been accumulated to a very considerable extent, and a commencement made of the determination of their equivalent forces" (quoted by Thomas, 1991, p. 67). And by the time he had carried out a series of experiments on the grand synthesis of electricity and gravity, all leading to negative results, he could still say in 1849 that they "do not shake my strong feeling [here is his informal introduction of personal views and biases into his methodology] of the existence of a relation between gravity and electricity" (Jones, 1870, p. 253).

There is a clear statement of prior beliefs (beliefs held prior to taking any data that might support those beliefs) in these quotes. We know that Faraday persevered in experimentation in the face of results disconfirming his strong prior beliefs, with eventual success in the case of magnetism generating electricity, but without it in the case of the relation between gravity and electricity. Nevertheless, Faraday himself believed the myth that science is strictly objective, arguing against excessive theorizing in science that when observations are not taken at face value but are interpreted through theory, they are necessarily prone to distortion (Gooding, 1985). In *The Inertia of Mind* (1818), Faraday warned of the dangers of becoming prematurely attached to one's own theories and looking only for confirmatory evidence. But he also sees the value in such attachment, and for the same reason, that it encourages the search for confirmation.

Faraday's belief in the unity of forces went even further. He speculated that electromagnetic waves analogous to those in the ocean exist, but he made the speculation in a sealed letter to the Royal Society which by his orders was not opened until 1937. On this issue, Faraday counted on future generations to collect the data to support his prior belief. [Albert Einstein (see Section 4.13) and other scientists did very much the same thing.]

REFERENCES

Agassi, Joseph (1971). *Faraday as Natural Philosopher*. Chicago: University of Chicago Press.

Berman, Morris (1978). *Social Change and Scientific Organization: The Royal Institution, 1799–1844*. Ithaca, NY: Cornell University Press.

Cantor, Geoffrey N. (1985). "Reading the Book of Nature: The Relation Between Faraday's Religion and His Science," pp. 69–81 in David Gooding and Frank A. J. L. James (Eds.), *Faraday Rediscovered*. Bassingstoke, Hants, UK: Macmillan. (Published in the United States and Canada by Stockton Press, New York.)

Faraday, Michael (1816a). *Chemistry Lectures*. London: Institute of Engineers.

———(1816b). "Observations on the Inertia of the Mind," in *Commonplace Book, 1816–1846*. London: Institute of Engineers.

———(1839–1855). *Experimental Researches in Electricity*. Reprinted from the *Philosophical Transactions* of 1831–1852 with other electrical papers from the *Quarterly Journal of Science, Philosophical Magazine, Proceedings of the Royal Institution*, 3 vols. London.

———(1854). "Observations on Mental Education," pp. 39–88 in *Lectures on Educa-*

tion Delivered at the Royal Institution of Great Britain. London: J. W. Parker and Son. Reprinted in *Experimental Researches in Chemistry and Physics* (1859). London: R. Taylor and W. Francis. Page numbers refer to reprinted edition.

Gooding, David (1985). "'In Nature's School': Faraday as an Experimentalist," pp. 105–135 in David Gooding and Frank A. J. L. James (Eds.), *Faraday Rediscovered*. Bassingstoke, Hants, UK: Macmillan. (Published in the United States and Canada by Stockton Press, New York.)

Gooding, David, and James, Frank A. L. J. (1985). *Faraday Rediscovered: Essays on the Life and Work of Michael Faraday, 1791–1867*. Bassingstoke, Hants, UK: Macmillan. (Published in the United States and Canada by Stockton Press, New York.)

Jones, Bence (1870). *The Life and Letters of Faraday*, 2 vols. London: Longmans, Green.

MacDonald, D. K. C. (1964). *Faraday, Maxwell, and Kelvin*. Garden City, NY: Doubleday.

Miller, Mabel (1968). *Michael Faraday and the Dynamo*. Philadelphia: Chilton.

Thomas, John Meurig (1991). *Michael Faraday and the Royal Institution (The Genius of Man and Place)*. Bristol, UK: Adam Hilger.

Tricker, R. A. R. (1966). *The Contributions of Faraday and Maxwell to Electrical Science*. Oxford: Pergamon Press.

Williams, L. Pearce (1965). *Michael Faraday: A Biography*. London: Chapman & Hall.

———(1966). *The Origins of Field Theory*. New York: Random House.

4.9   CHARLES DARWIN (1809–1882)

Charles Darwin was an English naturalist who demonstrated that the various species of life, both plants and animals, have been evolving steadily over time rather than having been created all at one time. This biological principle is called the *theory of evolution*. Darwin proposed that the mechanism by which this evolution takes place is that of *natural selection*.

4.9.A   Brief Biographical Sketch

Darwin attended Edinburgh University with the intention of studying medicine; but he was a poor student. What he did enjoy was collecting marine animals in tidal pools, accompanying fishermen trawling for oysters, and learning to skin and stuff birds. Since he had so little interest in medicine, he left Edinburgh at 18 and was sent to Cambridge to prepare for Holy Orders in the Church of England. Although he paid little attention to his official studies, he became acquainted with some distinguished scientists, in particular, John Stevens Henslow, reverend and professor of botany, who influenced Darwin profoundly by stimulating his interest in natural history.

In 1831, when Darwin was just 22, the Admiralty asked for a naturalist to accompany Capt. Robert Fitzroy of the Royal Navy on a voyage in the ship HMS *Beagle* to survey the coasts of Patagonia, Tierra del Fuego, Chile, and Peru, to visit some Pacific islands, and to establish a chain of chronometrical stations around the world. Henslow recommended Darwin. Darwin happily agreed to go and set sail from England on the *Beagle* on December 27, 1831. He was to be away for five years gathering data that would change forever the world's thinking about the origin of species. He took with him a copy of the first volume of Lyell's new book, *Principles of Geology*. He was about to begin emulating his hero, the great German naturalist Alexander von Humboldt (von Humboldt, 1769–1859, is treated in detail in Section 4.7), who traveled the world recording data on geology, plants, physical geography, oceanography, and many other features of the earth. Von Humboldt recorded his scientific observations in the five volumes of his tome *Kosmos* and drew theoretical generalizations from that systematization. He did not use these observations, as Darwin was to do, to derive a grand theory of the origin of species.

The Cape Verde Islands provided Darwin with his first observation of a volcano; in Brazil he saw his first tropical forest; in Argentina he found his first fossils: sloths, mastodons, and horses. In Chile, Darwin experienced an earthquake and observed both its effects in raising the level of the land and its connection with volcanic eruption (see Darwin and Seward, 1903, p. 95). Repeatedly when ashore he went on long, arduous, and dangerous expeditions on horseback, collecting specimens. Wherever he saw a mountain he climbed it, and on one journey from Chile to Argentina over high passes of the Andes, he was bitten massively by insects. After visiting the Galápagos Islands, the *Beagle* sailed to Tahiti, New Zealand, Australia, Cocos Keeling Atoll, Mauritius, South Africa, St. Helena, Ascension Island, Brazil again (to check chronometers), and then home. Darwin returned to England, on October 2, 1836, at 27 years of age.

The young man had, during all the hardships and dangers of the voyage of the *Beagle* and his numerous hazardous journeys ashore, enjoyed robust health and great physical stamina. But within a few months of his return began to show increasingly frequent symptoms of illness that reduced him to a state of semi-invalidism. These symptoms were great lassitude, painful intestinal discomfort, frequent vomiting, and sleeplessness. His doctors were unable to find any organic cause for his condition, in which he settled down to a daily routine of four hours' work, walks in the garden, and rests on a sofa smoking a cigarette while being read to. Frequently, he was unable to work at all.

His illness was never satisfactorily diagnosed. One major hypothesis was that it was psychological, relating to some sort of rebellion against his father. Another explanation proposed was his bites from insects: the massive attack that Darwin suffered in 1835 came from the bites of *Triatoma infestans*, the most important carrier of the trypanosome of Chagas' disease. This trypanosome, not discovered until 1909, is found in the blood even many years after infection; it causes lassitude and heart block and prevents normal functioning of the intestines. The case histories of patients with Chagas' disease and Darwin's symptoms fit like a glove: He suffered a heart attack in 1873 and died of another in 1882 at 73. It is against this background of suffering that Darwin's scientific work was achieved and that his family life was lived. Ten children were born, of whom two died in infancy.

4.9.B Darwin's Scientific Contributions

Much of Darwin's work in evolutionary biology and geology stemmed directly from the observations and collections that he made during the voyage of the *Beagle*. Charles Darwin wrote 15 books, plus four monographs on the taxonomy of barnacles. Most of these works were of minor importance relative to his two most important: *Origin of Species* (1859) and *Descent of Man* (1871). In the eyes of posterity, the other 13 books were so overshadowed by Darwin's momentous work on evolution that they have been neglected, but they were fundamental to his later work.

Starting from the fact that coral polyps are found only in clear salt water less than 20 fathoms deep, at temperatures not less than 68°F, and that coral atolls and barrier reefs are all at about sea level, Darwin argued that such atolls and reefs could only have resulted from subsidence of the sea floor, the corals growing upward as their bases dropped. Darwin's view has since been confirmed by deep borings of coral reefs revealing at depths of nearly 5000 feet dead corals that once lived within 120 feet of the surface. So Darwin was not only a keen observer of his surroundings, he also took the next (scientific) step of hypothesizing about how to integrate those observations into a unifying theory.

Another kind of reef, which Darwin called the *fringing reef*, occurs above sea level and results from elevation of the sea floor. Plotting on the map the distribution of atolls and barrier reefs on the one hand and of fringing reefs on the other, Darwin saw that great areas of the ocean bottom had undergone subsidence, and others elevation, and that all active volcanoes are in the latter. This agreed with the association that he had observed in South America between volcanic action and elevation of the ground. That such changes of level could be substantial he knew from his explorations in the Andes. There, at an altitude of 7000 feet, he had discovered a fossil forest overlain by thousands of feet of sedimentary deposits laid down by the sea, thus proving the occurrence of earlier earth movements of the order of 10,000 feet vertical height.

In petrology his comparison of volcanic lavas with plutonic rocks (rocks that have originated deep within the earth) showed that they were closely related (see also our discussion of the scientific work of Alexander von Humbolt in Section 4.7). The minerals in crystalline granites and in glasslike lavas were similar. He studied the direction of the veins of rocks containing the minerals and the angle in a vertical plane of the mineral rock strata. He showed that planes of cleavage were constant over very wide areas, parallel to the direction of great axes along which elevation of land had taken place over hundreds of miles. Furthermore, these planes of cleavage had no relation to the planes of stratification of sedimentary deposits, and had been superimposed on strata by pressure and recrystallization. This was the origin of the *deformation theory* of metamorphic rocks. His work in geology was to play a major role in Darwin's approach to science.

But Darwin's contribution to biological science was of a different nature. When he started on the voyage, like everyone else, he did not question the immutability of species. During the journey, however, his own observations raised several questions that set him thinking. Why do so many similar animals exist so far apart geographically? Why does the South American rhea, for example, resemble so closely the African ostrich? On the other hand, why are adjacent areas populated by similar, though not identical species? Why, for instance, were the birds and tortoises of each Galápagos island different, although the physical conditions of the islands seemed identical?

After his return to England, Darwin saw, in 1837, that these questions and many more, in comparative anatomy, embryology, classification, geographical distribution, and paleontology, could be explained satisfactorily if species were not immutable but had evolved into other species, many with a common ancestor. The evolutionary view of how humans descended by mutating over the generations, appears to answer all these questions, which otherwise remain inexplicable and without a common determining principle.

Darwin realized that it would be useless, in the state of opinion of his day (opinion very much dominated by church dogma), to try to convince anybody of the truth of evolution unless he could also explain how it was brought about. In searching for this cause, he knew that the key to human success in producing change in cultivated plants and domestic animals was careful selection of parents from which to breed the desired qualities, and he felt sure that selection must somehow also be operative in nature's creation of species. He knew that all individuals in a species were not identical but showed variation, and he realized that some individuals, well adapted to the places they occupied in the economy of nature (in the mid-twentieth century called *ecological niches*), would flourish, while others, less adapted, would perish. This was the principle of natural selection that he had grasped as early as 1837, but he still needed to understand how nature enforced it.

In 1838, Darwin read Malthus's *Essay on the Principle of Population*. In this work Malthus tried to show that because the rate of increase of human population was in a geometrical ratio while that of increase in human food supply was only in arithmetical ratio, the result must be misery and death for the poor, unless population growth was checked. Malthus's argument was unsound because it has never been determined to what extent human food supply could be artificially increased if it were given sufficient priority and finance. But Darwin saw at once that this fallacious argument could be applied correctly to plants and animals, which are unable to increase their food supply artificially. He saw, too, that in these organisms, mortality must be very high, thereby automatically enforcing the mechanism of selection of parents of successive generations. The note, in telegraphic style, which Darwin entered that day in his *Notebook on Transmutation of Species* stated: "On an average every species must have same number killed year with year by hawks, by cold, & c.—even one species of hawk decreasing in number must affect instantaneously all the rest. The final cause of all this wedging must be to sort out proper structure. . . . One may say there is a force like a hundred thousand wedges trying to force every kind of adapted structure into the gaps in the economy of nature, or rather forming gaps by thrusting out weaker ones." In these words, Darwin showed that he had solved the problem of the origin and improvement of adaptation as a result of selection pressure, and that modification during descent (i.e., evolution) does not take place in a vacuum but is strongly related to the ecological niches occupied by the species. Darwin is therefore included among the founding fathers of the science of ecology.

Having discovered the greatest general principle in biology, Darwin kept it to himself. In 1842 he penciled a "Sketch" of his results, which he expanded in an "Essay" in 1844, but he showed it only to his botanist friend Joseph Dalton Hooker. From 1846 to 1854 he devoted his attention and energies to a study of the different species of living and fossil barnacles, for the purpose of classifying them. This tedious work provided him with firsthand experience of the amount of variation found in species and of the problems of classification, essential for the study of how species originate. In 1856, Darwin started to put on paper his discoveries about evolution and natural selection. As he worked he added to his evidence by studies on the problems of divergence (i.e., the greater variability of species belonging to wide-ranging genera containing many species) and on geographical distribution, including the function of sea and wind in disseminating the population of oceanic islands. He also worked out his ideas in discussions with his friends Lyell, Hooker, and Thomas Henry Huxley.

Darwin's writing work went steadily until, on June 18, 1858, out of the blue, he received from Alfred Russel Wallace, a naturalist then in the Malay Archipelago, a succinct but complete statement of his own conclusions on evolution and natural selection. Darwin's shock at the danger of being forestalled in work on which he had been engaged for 20 years was great. But Lyell and Hooker, insisting that a joint paper by Darwin and Wallace should be read before the Linnean Society of London on July 1, 1858, saved the situation. Darwin then started what he called an *abstract* of the full work on which he was engaged. By 1859, the scientific atmosphere surrounding the origin of human beings and their history was saturated with the possibility of evolution. It was only a matter of time before someone published a systematic theory.

Darwin's abstract was the *Origin of Species*, which was published on November 24, 1859, 23 years after he had collected the data for his theories during the voyages of HMS *Beagle*; the book sold out immediately. By January

1872 the work had run through six editions. With the publication of the *Origin of Species*, Darwin brought down on himself enemies of two kinds. The first were old-fashioned scientists, some like Adam Sedgwick refusing to admit Darwin's method of using hypotheses as acceptable in science. Another was Richard Owen, who had until then enjoyed the reputation of being the leading English biologist and who was jealous, realizing that his former friend Darwin might eclipse him altogether and must, at any cost, be discredited. The other class of enemies were the upholders of orthodox religious beliefs, to whom Darwin administered two shocks. If evolution were true, the account of the Creation in the Book of Genesis was false or at least not literally true. Also, if evolution worked automatically by natural selection, there was no room for divine guidance and design in the production of living plants and animals, including humans, on earth.

The battle was joined at the Oxford meeting of the British Association for the Advancement of Science on June 30, 1860. Owen had carefully coached Samuel Wilberforce, Bishop of Oxford, who attacked Thomas Henry Huxley. Huxley came to be referred to as Darwin's "bulldog," because he acted as Darwin's principal defender in public debates about the controversial new theory. Wilberforce spoke in a patronizing and contemptuous manner to Huxley about Darwin's views (by referring to Huxley's simian origins). Huxley's counterattack was so brilliant that the Church of England never again formally attempted to cross swords with science. But there are many in society at large who have, to this day, refused to accept Darwin's theory of human evolutionary descent, preferring instead a literal interpretation of biblical sources.

4.9.C Major Works

The complete works of Charles Darwin may be found in the Cambridge University Library. A good source for publication of his works is John Murray, London.

Darwin, Charles (1832–1836). *Journal of Researches into the Geology and Natural History of the Various Countries Visited by H.M.S. Beagle*; popular title, *The Voyage of the Beagle* (1839). London: John Murray.

———(1842). *The Structure and Distribution of Coral Reefs* (2nd ed., 1874). London: Smith, Elder.

———(1859). *On the Origin of Species by Means of Natural Selection, or the Preservation of Favoured Races in the Struggle for Life*. London: John Murray.

———(1862). *The Various Contrivances by Which Orchids Are Fertilised by Insects*. London: W. Pickering, 1988; Series title: Darwin, Charles (1809–1882), Works of Charles Darwin, 1986.

———(1865). *The Movements and Habits of Climbing Plants*. London: John Murray.

———(1868). *The Variation of Animals and Plants Under Domestication*, 2 vols. New York: D. Judd.

———(1871). *The Descent of Man, and Selection in Relation to Sex*, 2 vols. London: John Murray.

———(1872). *The Expression of the Emotions in Man and Animals.* London: John Murray.

———(1875). *The Insectivorous Plants.* London: John Murray.

———(1876). *The Effects of Cross and Self and Cross Fertilisation in the Vegetable Kingdom.* London: John Murray.

———(1877). *The Different Forms of Flowers on Plants of the Same Species.* London: John Murray.

———(1880). *The Power of Movement in Plants.* London: John Murray.

———(1881). *The Formation of Vegetable Mould Through the Action of Worms.* London: John Murray.

4.9.D Subjectivity in the Work of Darwin

In Darwin's day science was supposed to make progress only by inductive methods. Although Darwin felt that he followed Baconian principles, and without any theory collected facts on a wholesale scale, he himself invalidated this when he wrote to Lyell (June 1, 1860): "Without the making of theories, I am convinced there would be no observations." In 1861 he wrote to a Cambridge political economist, Henry Fawcett: "About thirty years ago there was much talk that geologists ought only to observe and not theorise. . . . How odd it is that anyone should not see that all observation must be for or against some view, if it is to be of any service!"

Again in 1863, Darwin wrote to an Edinburgh botanist, John Scott: "*Let theory guide your observations*, but till your reputation is well established, be sparing in publishing theory. It makes persons doubt your observations" (*More Letters . . .* , vol. 2, p. 323). Thus, Darwin recognized that there is no such thing as pure induction from observations, for if the observer did not already have an idea of what to look for, a subjective personal belief, derived from deduction, nothing at all could be observed. We also find here a partial explanation for why Darwin waited 23 years to publish his findings on evolution through natural selection.

Darwin's methodology was to spin a hypothesis about anything that struck his attention (i.e., anything that he was predisposed by ideas to see), and then to deduce from it consequences that should follow and could be refuted or verified. This "hypothetic–deductive" method, as Sir Peter Medawar has called it, is illustrated in Darwin's letter to F. W. Hutton (Apr. 20, 1861): "I am actually weary of telling people that I do not pretend to adduce direct evidence of one species changing into another, but I believe that this view is in the main correct, because so many phenomena can thus be grouped and explained."

All of Darwin's mental energy was focused on his subject, and that was why poetry, pictures, and music ceased in his mature life to afford him the pleasure that they had given him in his earlier days. His technique for collecting facts, in addition to those that he observed for himself, was illustrative of tenacity and his capacity for holding strong personal beliefs. Darwin held some rather naive views. To Thomas Huxley he wrote (Jan. 9, 1860): "The history of error is quite unimportant." But the study of errors of measurement, and the study of errors that represent imperfections in a mathematical or statistical model, have become the focus of entire fields of study since Darwin's time.

Darwin's first transmutation notebook compiled aboard the *Beagle* was begun about July 1837. At that time Darwin was thinking about changes in organisms. He wrote that changes in the physical environment provoke and require changes in the organisms that inhabit that environment. At the time, people still believed in spontaneous generation of organic life from inorganic materials. (Although other scientists had earlier cast doubt on this theory, it was only later that Louis Pasteur showed that such a mechanism could not hold true; see the essay on Louis Pasteur in Section 4.10 for more details.) So it was that Darwin developed the theory of *monadism*, that simple living particles, or "monads" are constantly springing into life. They have their origin in inanimate matter and are produced by natural forces. Such was his subjective interpretation of the data as he saw the matter at that time. It was only later that he discarded this subjective interpretation of the same data.

Charles Darwin's research, beliefs, theories, and general scientific laws were summarized in his many published books, articles, and letters. But much of his writing was based on the notebooks he created during his five-year voyage on the *Beagle*. Those notebooks he casually entitled, B, C, D, E, M, N, Journal, and some "Old and Useless Notes." The M and N notebooks (written 1838–1839) were published for the first time only in 1974, in a book by Howard Gruber and Paul Barrett. The M and N notebooks focused on man, mind, and materialism as they related to Darwin's thinking about evolution. They bear on Darwin as a psychologist, because they treat such topics as memory; psychopathology; heredity, environment, and free will; beauty and imagination; music and poetry; instinct versus intellect; God's will, man's will, and chance; mind and body; emotion in humans and animals; dreams and thought; daydreams, dreams, and belief; happiness; passion; and many other topics normally confined to the realms of psychology, philosophy, and theology. Gruber, a psychologist, and Barrett, a biologist, were a very natural team to study these previously unexamined notebooks.

To understand the development of the theory of evolution in Darwin's mind, we need to understand that in his time, beliefs about evolutionary biology were shaped, in part, by philosophy developed by the Greeks. For example, according to Plato's notion of *ideal forms*, or *essentialism*, the physical world was a mirage from which little reliable information could be gained. The only things that really existed were changeless ideas or forms; the objects that existed in the physical world were distorted changeable shadows of these permanent unalterable essences. So change and variation were mere illusions, and genuine reality consisted of fixed types, permanently distinguished from one another. If this notion were applied to nature, it would rule out the possibility of biological change.

Evolution had been advocated earlier. J. B. A. P. de Monet Lamarck (1744–1829) developed the theory that new organs arise from new needs, that new organs continue to develop in proportion to the extent to which they are used, and that these acquisitions are handed down from one generation to the next. Conversely, disuse of existing organs leads to their gradual disappearance, and these shrinking organs would also be inherited. Lamarck suggested that living matter had a natural ambition to be bigger and more complex. But his neglecting to provide evidence for evolution, his attempt to explain its cause by appealing to the "inner feeling" of a tendency to perfection (an example of the use of strong subjectivity and a strong belief in science regardless of the availability of supporting data), and his equally fanciful belief in the satisfaction of the needs of the organisms led other scientists to reject his evolutionary theory.

Erasmus Darwin (1731–1802), Charles Darwin's grandfather, was a distinguished doctor and evolutionary biologist who believed in the "descent of man through modification." He had his own ideas about evolution, similar to those of Lamarck. Drawing conclusions from simple observations, Erasmus Darwin believed that species changed because they needed to adapt to their environment.

The French naturalist Comte G. L. L. de Buffon (1707–1788) published his *Theory of the Earth* in 1749. In this work he took an alternative path to the one in the Bible (Creationism, and belief in the assertion of Archbishop James Ussher in 1650 that the universe was created on October 22, 4004 B.C.E., making it about 6000 years old). Buffon declared that in his opinion the earth was very old, that clerical people had seriously underestimated its age, and that living things had probably undergone substantial change. Moreover, to make his point, he identified much of the evidence later presented by Charles Darwin. He attempted to order chronologically the appearance on earth of different species. In essence, he was supporting the doctrine of descent with modification.

In 1844, the Scottish publisher, Robert Chambers (anonymously), published the book *The Vestiges of the Natural History of the Creation*. The book suggested that the succession of fossil types was evidence of an unceasing transformation of what God had created at the beginning of time. The book was considered blasphemous. However, Charles Darwin felt that the book expressed many of his own views, and therefore he postponed until 1859 publication of his own theories in what later was to become *Origin of Species*. By that time the scientific world was very ready to accept what he said.

All of this earlier evolutionary biology was well known to Charles Darwin as it was to other scientists. Moreover, Darwin's observational data collected during his voyages on the *Beagle* were carefully presented to the scientific world. Thus because all of these observational data were then available to the biological science community, as an augmentation of natural history data collected earlier, other scientists could use the same data set as Darwin to develop their own theses regarding the descent of man. One needs, then, to ask why it was Darwin and not one of his contemporaries who published the definitive work on the origin of species.

During the 23 years between his return from South America until the publication of *Origin of Species*, Darwin introspected about the meaning of his observational data, about how all the loose ends might fit together most logically, and about how earlier theory might be brought to bear on the patterns of evolution of living things and a dynamically changing earth. This introspection process involved a subjective synthesis of observational data with strong personal beliefs about the nature of the patterns from which the observational data had come. His interpretation of the data, as distinguished from the interpretations of them by other scientists, depended on his own background, his own subjective evaluation, his creativity, and his own formulation of scientific hypotheses that could be tested. Because his beliefs differed from those of the earlier biologists, he saw different patterns in the data than they did. Thus, in light of both his strongly and carefully considered beliefs, and his observational data, his (posterior) inferences about evolution and the descent of man were different from theirs.

To develop his ideas about the descent of man, Darwin had to free himself from earlier theories supporting a constancy in nature stemming from a Creationist, Bible-based, static earth, as well as a constancy of all living species. Then he had to construct alternative hypotheses regarding the mechanisms by which humans have developed into our present form, and these mechanisms were not always strongly supported by large quantities of observational data. Yet Darwin's powerful insights, intuition, lengthy observations of nature, and ability to recognize patterns where others could not, generated within him very strong personal and subjective beliefs about human evolution. It was these strong subjective beliefs that propelled him forward to promulgate his theses, despite strong opposition in the lay community, opposition that exists to the present time. Fortunately, he did not have to battle much with the scientific community as well; by the time he published *Origin of Species*, the scientific community was generally quite willing to accept these ideas, and it largely felt that the ideas were long overdue in coming. It was Darwin's subjective views of the changing, evolutionary process in nature that dominated his theories about human descent.

But there were a few important objections from scientists, and these objections led Darwin to a totally subjectively based, remarkable recanting of his earlier notions about the heritability of acquired traits (his *theory of Pangenesis*, described below). There were four specific objections that gave him the most trouble. First, the zoologist H. St. George Mivart argued that although natural selection might account for the success of well-established adaptations, it could not possibly explain the initial stages of their development. The biological usefulness of the eye is self-evident, but how did such an organ get started in the first place? We now recognize that Darwin's explanation is true; that at its first appearance, a fortuitous novelty may confer subtle and invisible advantages. Nevertheless, the utility of imperceptible novelties continued to be a problem and was one of the reasons why natural selection became discredited in Darwin's own lifetime. (Although the theory of evolution continued to be accepted, the mechanism of natural selection was rejected in favor of Lamarckian principles.)

Second, Darwin was well aware of the fact that there were huge gaps in the fossil record, so that direct evidence of smooth transitions from form to form was not available. He explained these by assuming subjectively that the intermediate stages had been destroyed. To him, it was as if some geological vandal had torn out pages and chapters in the book of life. But he maintained strong beliefs with no evidence (prior beliefs) that subsequent research would restore these lost episodes and that the continuity of the record would eventually be restored. This has not happened. Modern paleontologists now recognize that Darwin did not explain the often-abrupt succession of fossil types. There is now overwhelming evidence pointing to the conclusion that certain forms remained stable for long periods of time, only to be succeeded suddenly by new forms altogether. Latter-day Creationists have seized on this finding, in the effort to reinstate the doctrine of serial creations. Serious biologists dismiss this as a frivolous suggestion. But they are ready to concede that the process of evolution is more episodic than Darwin supposed. Confronted by the fact of so many unbridgeable gaps in the fossil record, paleontologists are trying to accommodate themselves to the idea that the modification that accompanies descent is not necessarily gradual. Although the process of imperceptible change has an all-important part to play in the origin of species, it is often superseded by abrupt transformations that result in the emergence of comprehensively new designs.

The exact mechanism of these rapid changes is now a subject of heated controversy. One biologist has described these novelties as "hopeful monsters" and has argued that an unsolicited transformation in the genetic instructions might occasionally result in the emergence of a design fortuitously adapted to the new circumstances, thrown up at a time of rapid geological change. Although this suggestion kills two birds with one stone, since it deals both with the apparent absence of intermediate types and with the apparent uselessness of incipient novelties, it presupposes disruptively large changes in the hereditary instructions. Modern geneticists argue that such abrupt upheavals would prove lethally disabling to the process of embryological development. Whatever the explanation is, it must eventually account for the fact that evolution has not always been a smooth process and that the history of life on Earth has often been sharply punctuated.

The third objection to Darwin's theory was that an evolutionary theory based on the slow accumulation of small invisible novelties presupposes huge lengths of time. As we have already seen, the geologists of the eighteenth and nineteenth centuries had gradually recognized the great age of Earth. But Darwin's theory demanded an almost inconceivable span of biological time. This assumption was seriously endangered when the physicist Lord Kelvin (incorrectly) calculated from the temperature of Earth's interior that Darwin had grossly overestimated the age of the globe. Darwin's subjective belief, which was borne out in time, was that Kelvin's calculation would turn out to be wrong. If he had lived longer, he would have been gratified to discover that Earth was even older than he himself supposed.

Darwin's ignorance about the mechanism of inheritance left him exposed to a fourth and potentially even more damaging objection. In 1867, a Scottish engineer, Fleeming Jenkin, pointed out that a favorable variation would soon disperse itself as the "fortunate" individual interbred with "normal" members of a population. Jenkin's objection (later proved false) was based on the assumption that the genetic factors were infinitely divisible, which implied that a new variation would automatically distribute itself in steadily diminishing amounts. Ironically, this objection could have been answered, if the scientific world had taken note of a discovery published by the Czech monk Gregor Mendel (see Section 3.3 for further discussion of Mendel) less than a year after Jenkin's assertion. Working in the obscurity of a provincial monastery, Mendel proved that the genetic factors behaved as if they were indivisible particles and that they did not blend or dilute themselves in the course of interbreeding. Unfortuntely, Mendel's paper was neglected, and by the time it was rediscovered in 1900, the theory of small unsolicited novelties had undergone a total eclipse.

Confronted by the twin specters of Kelvin's calculation of the age of Earth and Jenkin's prediction of the dispersion of fortunate variations, Darwin began to lose confidence in the effectiveness of natural selection. He now felt it necessary to introduce some auxiliary process that would hasten evolutionary change in a purposive direction. In summarizing the sixth and final edition of *Origin of Species* (1872), he conceded that evolution was ". . . aided in an important manner by the use and disuse of parts; and in an unimportant manner, that is in relation to adaptive structures, whether past or present, by the direct action of external conditions. . . . It appears that I formerly underrated the frequency and value of these latter forms of variation, as leading to permanent modifications of structure independently of natural selection." By introducing this codicil, Darwin reverted to his original belief in the Lamarckian heredity effect of effort and experience. In 1868 he published a two-volume book containing an elaborate theory (pangenesis) that purported to explain the inheritance of acquired characteristics (*The Variation of Animals and Plants Under Domestication*). In his theory of pangenesis, Darwin formulated a set of ideas that were very similar to ideas put forward by the ancient Greek philosopher Democritus. According to this theory, the cells destined to play a reproductive role gradually accumulate a set of representative particles, or "gemmules," derived from all the organs and tissues of the adult body. It was as if the body were divided into a series of parliamentary constituencies, each of which sent a team of representatives commissioned to reproduce the limbs, organs, or tissues for which they stood. These gemmules were dispatched into the bloodstream; and having reassembled in the reproductive cells, their presence guaranteed a faithful reduplication of the parent physique.

For Darwin, this theory proved an admirable explanation for the inheritance of acquired characteristics. If, through its own efforts, a creature succeeded in enlarging the muscles of its limbs, the number of gemmules arising from these overgrown parts would increase, and they would therefore be overrepresented in the constituent assembly gathered in the reproductive cells. The offspring would therefore automatically inherit the fruits of its parents' exertion. Conversely, if a limb or organ dwindled through disuse, the gemmules would be underrepresented and the offspring would inherit the parental deficit. This revision gave heart to those who already found the evolutionary role of unsolicited variation suspect; and the fact that Darwin himself had made such a concession was one of the factors that led to the development of a Lamarckian backlash. From 1870 the revision gathered momentum, and in the years shortly before Darwin's death there was a veritable landslide in favor of Lamarck.

In Europe and in the United States, evolution was widely accepted as a doctrine, but natural selection was rejected in favor of the inherited effects of use and disuse, and the hereditary effects stimulated directly by the environment. By the end of the century, only two important scientists were prepared to regard unsolicited variation as the raw material of evolutionary change. Both of them remained unswervingly loyal to the principle of natural selection: Alfred Russell Wallace and the German naturalist August Weismann. After sifting all the experimental evidence, Weismann was unable to find a single case in which the experience and effort of one generation had influenced the structure and function of the next. Experimental scars and mutilations were never inherited; nor were skills. The brawny muscles acquired by a blacksmith during a lifetime of toil were not handed on to his idle son. Regardless of the changes undergone during the lifetime of any one individual, the next generation invariably reverted to type. According to Weismann, the doctrine of inherited characteristics was factually wrong and biologically impossible. The scientific community eventually abandoned pangenesis.

These various contradictions were mostly reconciled, in time. Lamarckian explanations were ruled out when it was shown that acquired characters, whether impressed by the environment or resulting from use or individual effort, were never inherited. The advance of genetics showed that large mutations were rarer and of far less biological importance than those of small extent and that, apparently, continuous evolutionary change could be, and often was, brought about by the accumulation of numerous small discontinuous mutations under the guidance of natural selection.

Finally, R. A. Fisher in 1930 made it clear that heredity was particulate, dependent on distinct self-reproducing units or genes, each of which could mutate into new self-reproducing forms. Further, the fact that most mutants are recessive at once got rid of the major difficulties that beset Darwin, who accepted the current view of blending heredity—the view that characters and the entities that determined them were commingled into a single blend when crossed. Commingling would imply that any new character would be progressively diluted by crossing in each generation (as Jenkins had predicted), and would make its establishment in the stock difficult. But a particulate genetic mechanism in which most mutants are recessive makes it possible for new mutants to be stored indefinitely in the constitution, and for new combinations of new and old genes to be formed, ready to be utilized by selection when conditions are favorable. Natural selection was seen, not as involving the sharp alternatives of life or death, but as the result of the differential survival of variants; and it was established that even slight advantages, of one-half of 1% or less, could have important evolutionary effects.

Darwin spent many years traveling around the world gathering observational evidence for his theories, and then 23 years introspecting about evolution and the descent of man, and the mechanism for natural selection that drives it, before he elected to write about it (1836–1859). But as alluded to earlier, after *Origin of Species* went through five successful editions, the pressures of the scientific and general lay communities forced him to rethink his scientific beliefs in favor of an explanation that was not only totally subjective, but was also not based on any experimental data at all. This new explanation was in fact false, or so we now believe.

Darwin's subjective views were strongly influenced by observational data, both his own and that of earlier workers. Were any of these eighteenth- and nineteenth-century biologists mentioned above developing their scientific conclusions in an objective way? We don't think so. All of them, including Charles Darwin, started their scientific studies with differing subjective prior beliefs about the descent of man, that is, beliefs that were different in various respects from one another's. Their beliefs were shaped by their own backgrounds, experiences, understandings of earlier scientific thought on the subject, and differing degrees of resistance they were able to muster to counter the pressures of lay societal beliefs, political pressures, familial pressures, tacit promises of collegial ostracism if they were found wrong, church edicts, and problems of contradicting what was widely believed to be "common knowledge." Then, starting their scientific studies with these differing subjective prior beliefs, they each examined the available scientific observational data that was thought to bear on the subject. They then developed differing conclusions about what all the data meant; that is, their posterior inferences, not surprisingly, were sometimes quite disparate from one another. It is Charles Darwin's subjectively based (posterior) inferences, which depended on an informal mixture of belief, theory, and observational information, that have ultimately dominated scientific thought about natural selection and the origin of the species up to the present time.

REFERENCES

Darwin, Francis (Ed.) (1887). *Life and Letters of Charles Darwin*, 3 vols. London: John Murray.

Darwin, Francis, and A. C. Seward (Eds.) (1903). *More Letters: Letters of Charles Darwin*, 2 vols. London: John Murray.

De Beer, Sir Gavin (Ed.) (1974). *Darwin and Huxley: Autobiographies*. New York: Oxford University Press.

Encyclopaedia Britannica (1975). Vol. 5, pp. 492–496.

Fisher, R. A. (1958 [1930]). *The Genetical Theory of Natural Selection*. Oxford: Oxford University Press; New York: Dover.

Gruber, Howard E., and Paul H. Barrett (1974). *Darwin on Man*. Toronto: Clarke, Irwin.

Huxley, Sir Julian (1958). "Introduction," Mentor edition of *The Origin of Species by Charles Darwin*. New York: Mentor Books.

Miller, Johnathan, and Borin van Loon (1982). *Darwin for Beginners*. New York: Pantheon Books.

4.10   LOUIS PASTEUR (1822–1895)

Louis Pasteur was a French scientist whose world-famous research spanned the fields of crystallography, structural chemistry, bacteriology, microbiology, and the etiology of infectious diseases. The well-known process of *pasteurization* of milk and other dairy products is a purification procedure now required in most urban areas of the world.

4.10.A   Brief Biographical Sketch

Louis Pasteur was born December 27, 1822, in eastern France. His childhood was uneventful. At 18, he graduated with a baccalaureate degree in letters, and two years later received a second such degree, in science. After failing the entrance examination for the much-sought-after *Ecole Polytechnique*, an engineering-oriented school, the following year he was admitted to *l'Ecole Normale Supérieure*, the famous teachers' college in Paris. He received his doctor of sciences degree at 25, in 1847, in both physics and chemistry. A year later, he presented an important paper on crystallography before the Paris Academy of Sciences.

In 1848, Pasteur was appointed professor of physics at the Dijon *Lycee* (a secondary-level school). Months later he took a job at the University of Strasbourg as professor of chemistry. There, at 26, he married Marie Laurent, the daughter of the rector of the university. They had five children, only two of whom survived childhood. In 1854, when he was 32, Pasteur was appointed dean of the new science faculty at the University of Lille. Pasteur is reputed to have begun his studies on fermentation at Lille in order to respond to a question raised by an industrialist on the production of alcohol from grain and beet sugar. From studying the fermentation of alcohol he went on to the problem of fermentation in milk, showing yeast to be an organism capable of reproducing itself, even in artificial media, without free oxygen, a concept that became known as the *Pasteur effect.*

In 1857, Pasteur was appointed director of scientific studies at *l'Ecole Normale Supérieure.* Despite his additional administrative responsibilities, he continued his work on fermentation and concluded that fermentation was the result of the activity of minute organisms (yeast). Pasteur also showed that milk could be soured by purposely introducing a number of organisms from buttermilk or beer, but the milk would not sour if such organisms were excluded. These demonstrations ended a debate about the mechanism behind fermentation that had raged for many years. Pasteur was elected to the Academy of Sciences in 1862.

As a logical sequel to his work on fermentation (the conversion of sugar into alcohol and carbon dioxide in the presence of yeast bacteria), Pasteur began research on *spontaneous generation* (the concept that bacterial life arose spontaneously). This research continued into the mid-1860s. The question of life in the form of complex species arising from nothing was counter to Darwinian theory but consistent with theological doctrine, so there was a heated controversy over the matter among scientists of that period. (Charles Darwin published his *Origin of Species,* in English, in 1859; the French translation appeared in 1862; for details, see Darwin in Section 4.9). Pasteur recognized that both milk and alcohol fermentations took place more rapidly when there was exposure to air. So his research questioned whether invisible organisms were always present in the atmosphere or whether they were generated spontaneously. Pasteur showed experimentally that food decomposes when there are microbes present in the air that cause its putrefaction. Moreover, food does not putrefy in such a way as to generate new organisms within itself spontaneously.

In July 1867, when Pasteur was almost 45 years old, an important incident involving a student affected his life profoundly. The student had written a letter supporting a speech by a French senator that had deplored an attempt to remove allegedly subversive books from a provincial library. The letter also approved of a recent attempt to assassinate Emperor Louis Napoleon. Pasteur recommended that the student be expelled and threatened to resign his university post unless the expulsion order were carried out. When the student was expelled, the entire student body went out into the streets of Paris to protest, closing down the school. By the beginning of the fall term in October, Pasteur, by then a very eminent scientist, had been dismissed from the university and the school had reopened.

In his teaching capacity, Pasteur was a government employee. Now that he had been dismissed from his university post, the government needed to find this renowned scientist a new job. In 1868, Pasteur was given both a professorship in chemistry at the Sorbonne and the directorship of a newly constructed, and generously endowed (by the Emperor Louis Napoleon) research laboratory in physiological chemistry. In 1871, Pasteur resigned from his teaching position at the Sorbonne so that he could devote himself more fully to research at the laboratory. In 1873, Pasteur was elected a member of the Academy of Medicine. In 1874 the French parliament provided him with a large stipend that was intended to provide him with financial security while he pursued his research.

In 1881, Pasteur succeeded in vaccinating a herd of sheep against the disease known as anthrax. He was also able to protect fowl from chicken cholera, for he had observed that once animals stricken with certain diseases had recovered, they were later immune to a fresh attack.

On April 27, 1882, at 60 years of age, Pasteur was elected a member of the Académie Française. He then began experiments designed to prevent the disease rabies. On July 6, 1885, he saved the life of a 9-year-old boy, Joseph Meister, who had been bitten by a rabid dog. The experiment was an outstanding success, opening the road to protection from a terrible disease. In 1888 the Pasteur Institute was inaugurated in Paris for the purpose of undertaking fundamental research, prevention, and treatment of rabies. Pasteur, although in failing health, headed the institute until his death on September 28, 1895.

4.10.B   Pasteur's Scientific Contributions

Louis Pasteur's scientific education and background prepared him to carry out research that traversed the borders of several of the physical and life sciences, including biology, chemistry, and physics. He began his career by studying some optical problems in crystallography. Then each new area of research he entered built logically and very efficiently on results from his previous research. In crystallography, Pasteur devoted the first 10 years of his scientific career (1847–1857) to studying the ability of *organic* substances to rotate the plane of polarized light shining on them and to studying the relationship of this property to crystal structure and molecular configuration. He examined two tartaric acids. The first, simply called *tartaric acid*, is found as a component of the tartar in wine fermentation vats. The second, known as *paratartaric acid* or *racemic acid* (which means "acid from grape") has the same chemical properties and structure as tartaric acid. But Pasteur discovered that the one acid (tartaric) turned the plane of a ray of polarized light to the right, while the other (racemic) had no corresponding optical effect. Pasteur set out to solve this mystery. He knew that *inorganic* substances such as quartz crystals have the ability to rotate a plane of polarized light passing through them (some to the right and some to the left). But at that time it was believed that the optical activity of organic substances was attributable to molecules in solution and had nothing to do with their crystalline structure. Nevertheless, Pasteur experimented with inorganic substances whose salt crystals turned the polarization plane to the left and with others that turned it to the right. Pasteur combined substances with the two types of crystals and found that the combination was optically inactive. By analogy, Pasteur abandoned the idea that optical activity of organic compounds was due to their molecular structure and concluded that it must be a consequence of crystalline structure, just as in inorganic materials. In particular, racemic acid must be composed of two isomeric acids whose crystals are mirror images of one another, which, when mixed in equal amounts, divert the ray, with equal power, in opposite directions. One of the two crystal forms of racemic acid proved to be identical with the tartaric acid of fermentation. This led to Pasteur's research into fermentation and his founding of the field of stereochemistry.

Pasteur also linked his research in crystallography to biology and life. He noted that the substances that seemed to exhibit crystalline asymmetry in solution (with mirror images of the crystals rotating the planes of polarized light in opposite directions) were organic, and those that were optically inactive in solution were inorganic and had symmetric crystalline structures. He argued that optically active substances were *naturally occurring organic compounds* (organic compounds are those that contain carbon), and they are necessary for life; organic compounds that are not active optically, and inorganic substances, are not necessary for life, so he considered them to be *secondary products*. It was from this vantage point that Pasteur attacked, and helped to destroy, the theory of spontaneous generation.

Fermentation was not yet understood in 1857. By 1865, Pasteur had shown that the various changes involved in the fermentation process were the result of the presence and growth of a microorganism which he called the *ferment*. He was thus the founder of the science of bacteriology (as well as that of stereochemistry). He proceeded to demonstrate that the varieties of fermentation observed were attributable to separate organisms in each case and that when these atmospheric germs were absolutely excluded, no change took place.

Pasteur also showed that one component of racemic acid (that identical with the tartaric acid from fermentation) could be utilized for nutrition by microorganisms, whereas the other, which is now termed its *optical antipode*, was not assimilable by living organisms. On the basis of these experiments, Pasteur elaborated his theory of molecular asymmetry. He showed that the biological properties of chemical substances depend not only on the nature of the atoms constituting their molecules (which were the same in the case of the two acids with which he had been experimenting), but also on the manner in which these atoms are arranged in space (which were different with his acids). In the course of his analysis he once again encountered—although in liquid form—new "right" and "left" compounds.

After laying the theoretical groundwork, Pasteur proceeded to apply his findings to the study of vinegar and wine, two commodities of great importance in the economy of France. He developed a sterilization process that came to be called *pasteurization*, the destruction of harmful germs by heat, which made it possible to produce, preserve, and transport these products without their undergoing deterioration. These discoveries proved to be of great value to the brewing and winemaking industries, where scientific certainty could now be substituted for the guesswork used earlier. Once the pasteurization process was introduced in France to prevent the deterioration of wines, beers, and milk, these industries became more profitable, and fewer consumers suffered ill effects from using spoiled products.

In 1865, Pasteur began the study of silkworm diseases; he researched the area for the next five years. At that time, silkworm diseases had tremendously reduced the value of raw silk in France. The entire industry was threatened. He carried out his research on silkworm diseases in the plantations of southern France. Eventually, he was able to determine the causes of the silkworm diseases and to eliminate them as a serious threat to the industry. His work was followed by a rapid increase in the annual value of silk and the reestablishment of prosperity in the trade.

From 1870 to about 1876, Pasteur devoted himself to the problems of beer going bad. Similar to wine and vinegar, beer was likely to undergo spontaneous changes, to acidify, and spoil, especially during the warm seasons. Pasteur demonstrated that these changes in the beer were always attributable to microscopic organisms. Following an investigation conducted both in France and among the brewers in London, he devised, as he had done for vinegar and for wine, a procedure for manufacturing beer that would prevent its deterioration with time. British exporters, whose ships at the time had to sail entirely around the African continent, were thus able to send British beer as far as India without fear of its deteriorating.

Smallpox was a disease that ravaged Europe and the Orient for many centuries starting from about the sixth century A.D. The disease affected its peak number of victims by about the seventeenth century. Those victims who lived (about 25% of those infected) were left with pockmarked faces and blindness in many cases. It was known that repeated attacks of smallpox were very rare. Once one recovered from it, one was quite unlikely to get it again. People began to understand that if they were purposely given (by injection) a small amount of the disease, as from a pustule, only a very mild case of smallpox would result, and the disease could then run its course with a minimum of risk. This approach to prophylaxis was called *inoculation.*

Edward Jenner had studied cowpox, a relatively harmless disease in the smallpox family which usually affects cows and people who work directly with them. He found, in 1796, that when a person stricken with cowpox was injected with virulent smallpox virus, the person would then not contract smallpox. Jenner had discovered an immunization for smallpox. This vaccination with cowpox virus immunized the person against the terrible scourge of the smallpox virus.

Pasteur's work in preventive medicine may be his best known achievement. This work started in about 1877 and consumed about the next two decades, until the end of his life in 1895. He developed a *germ theory of disease.* Pasteur generalized the work of Edward Jenner by applying similar principles of inoculation to fermentation and infectious diseases. Pasteur knew that each variety of fermentation was originated by a specific ferment; so in the case of various maladies, the infection depended on the presence of a specific microbe. He reasoned that an artificial growth of this microbe might be obtained and that by growing a certain succession of such cultures (by harvesting several microbe generations in the laboratory environment), the virus would be weakened in virulence. Then, if an animal were injected with such weakened strains of the virus, a slight attack of a given malady would appear, but the animal would then be rendered immune to the disease. Pasteur's application of this notion to cholera (bacilli) in chickens and to anthrax (bacilli) in cattle and sheep was very successful.

Pasteur finally extended the concept of immunization to inoculation for rabies virus (in 1885), with enormous success. After experimenting with inoculations of saliva from infected animals, he came to the conclusion that the virus was also present in the nerve centers. He demonstrated that, when a portion of the medulla oblongata of a rabid dog was injected into the body of a healthy animal, symptoms of rabies were produced. By further work on the dried tissues of infected animals and the effect of time and temperature on these tissues, he was able to obtain a weakened form of the virus that could be used for inoculation. Having detected the rabies virus by its effects on the nervous system and attenuated its virulence, he applied his procedure to human beings.

Louis Pasteur brought about a veritable revolution in disease understanding and treatment. By abandoning his laboratory and tackling the agent of the disease in its origin in the natural environment, he was able through his investigations to supply the complete solution to the question before him, not only identifying the agent responsible for a disease but also indicating the remedy. A skillful experimenter, endowed with a great curiosity and a remarkable gift of observation, Pasteur devoted himself with immense enthusiasm to science and its applications to medicine, agriculture, and industry.

4.10.C Major Works

Pasteur, Vallery-Radot (Ed. and Louis Pasteur's grandson) (1922–1939). *Oeuvres de Pasteur*, 7 volumes containing all books and papers published by Louis Pasteur during his lifetime, plus other works not published during his lifetime, and related documents by others), Paris: Masson et Cie.

———(1964). *Pasteur: Donation du Professeur*. Paris: Bibliotheque Nationale. (The donation of Pasteur's previously private laboratory notebooks to the National Library.)

———(1985). *Nouvelle Acquisitions Latines et Françaises du Departement des Manuscrits Pendant les Annes 1977–1982*: Inventaire Sommaire, Paris, 17923–18112. Louis Pasteur, *Papiers*.

4.10.D   Subjectivity in the Work of Pasteur

A half century ago, Dubos (1950) pointed out Pasteur's dependence on strongly held preexperiment beliefs, intuition, and deep understanding of the processes he worked with to advance his science: "In many instances, discovery appears to have evolved from the fact that Pasteur had been made alert to the recognition of a phenomenon, because he was convinced a priori of its existence" (p. 373). Or, again, "[h]e often mentioned his use of preconceived ideas, from which he derived the stimulus for many experiments" (p. 369). And yet again: "His greatest discoveries were the fruits of intuitive visions and they were published in the form of short preliminary notes long before experimental evidence was available to substantiate them" (p. 360).

More recently, careful study of Louis Pasteur's private notebooks has revealed that very broadly based informal subjectivity and personal beliefs pervaded much of his research methodology and his approach to scientific investigation. These revelations have caused a considerable stir in some scientific communities. We find, however, that they show that Pasteur was, like most other scientists, a research worker whose personal beliefs, based on an educated understanding of underlying scientific processes, assisted him in conjecturing what was probably true about the world. Often, but certainly not always, his conjecturing, intuition, and frequent use of educated strong personal belief turned out to be supported by his experimental results. We explore below some of the subjectivity issues associated with Pasteur's approach to scientific research.

It is important to make distinctions among the many ways other-than-objective methods are used by scientists. It will be seen in the discussion that follows that like other scientists, Louis Pasteur certainly used subjective research methods based on strongly held prior beliefs. But also, and very differently from some other scientists, some observers have asserted that Pasteur was much more than just subjective. They argue that he lied and misrepresented his "scientific" results—that he claimed to have carried out experiments that proved his beliefs when, in fact, no such experiments had been carried out. In short, Pasteur is being accused of scientific fraud. Fraud carries belief and subjectivity to extremes that are unacceptable in science, and a sharp line must be drawn between, on the one hand, the use of subjectivity to assist the scientist in conjecturing what might correspond to scientific truth, even approximately, and, on the other hand, fraud. It is mandatory for maintaining a proper perspective about science that such distinctions be borne in mind when evaluating claims made against Pasteur or any other scientist. Although some historians of science argue that such fraud was actually committed by Pasteur, others disagree.

Pasteur instructed his family that after his death his laboratory notebooks were not to be shown to anyone—ever. Despite this request, however, in 1964, his grandson, Pasteur Vallery-Radot, donated most of the collection to the

Paris Bibliothèque Nationale. By 1985, the Bibliothèque Nationale had prepared a printed catalogue of the notebooks and other works. Now that historians of science have examined this material, they have discovered large discrepancies between what Pasteur recorded in his notebooks, and what he did and said in public. Comparing the two, some have concluded that like other scientists, Pasteur was driven by very strong, preconceived beliefs in his own ability to discover and understand some of the truths in nature. Moreover, sometimes these strong beliefs were in disagreement with observed data, and sometimes, it seems, Pasteur failed even to report such discrepant data. His approach to scientific research was anything but the narrow, mythical path of objectivity that scientists are generally believed to take. His approach is described in detail in Geison (1995) and in some of Geison's earlier writing on Louis Pasteur (see, e.g., Geison, 1974).

Geison shows many discrepancies between what Pasteur wrote in the notebook and what he said in public or wrote for public consumption. He specifically denies any attempt to tarnish Pasteur's reputation, but believes that the telling of the story will shed light on the actual practice of science, showing that pure objectivity in science is really a myth. Instead, science, like any other form of culture, depends on rhetoric and the actions of human beings. Geison's 1995 book on Pasteur was reviewed by Lewis Wolpert (1995). Wolpert said: "Pasteur was deceptive because . . . he wanted both priority and recognition . . . to find support for his work and for personal gratification. . . . What this life of Pasteur shows is how complex, hard and imaginative scientific discovery is, and that it requires a variety of skills rarely found in one person."

Another review of Geison's 1995 book described the comparison of the notebooks with the published work and then had this to say about Pasteur's research methodology (Altman, 1995): "[T]his is not the only example of scientists and historians as well as investigative journalists beginning to shatter myths about crucial discoveries and those who made them. *The disclosures are revealing that science is not as objective, neat and scrupulously honest as it is portrayed*" [emphasis added].

Chapter 6 of Geison's book discusses the discrepancy between Pasteur's public and his laboratory-notebook accounts of the production of his vaccine against anthrax (which involved the "secret of Pouilly-le-Fort"). This discussion is primarily an account of the relationship between Pasteur and his rival Jean-Joseph Toussaint for claiming priority for discovering a vaccine against anthrax. The incident relates to Pasteur's claims to the scientific community and to the public about scientific results that were based on his generous use of his exceedingly powerful personal views about the underlying structure of the phenomenon he was studying, in this case, immunization against anthrax, prior to his obtaining the data required to justify his conclusions.

In the incident of Pouilly-le-Fort, Pasteur agreed to hold a public trial of a new vaccine he had discovered against anthrax, a major killer of animals and people. (The same anthrax microorganism is being discussed publicly these days as possibly being designed by some countries for use as a weapon for mass destruction of human beings in biological warfare.) The trials took place at Pouilly-le-Fort, a suburb of Paris, in May–June 1881. Twenty-five sheep were vaccinated with the Pasteur vaccine, and then these 25, along with another 25 unvaccinated sheep, were injected with virulent anthrax bacilli. With a very few exceptions that were explained away, the vaccinated sheep did very well compared with the unvaccinated sheep, most of which died. The public trials were a great success for Pasteur. Although Pasteur was never very clear about the actual nature of the vaccine he used in the trial, finally, on June 13, 1881, he announced to the French Académie des Sciences that his vaccine in the Pouilly-le-Fort trials was prepared by a method in which exposure to atmospheric oxygen played a crucial role. But in 1937, some 40 years after Pasteur's death, his nephew Adrien Loir published some essays (Loir, 1937–1938) in which he asserts that "the vaccine actually used at Pouilly-le-Fort had been prepared *not* by atmospheric attenuation, but rather by the 'antiseptic' action of potassium bichromate."

Geison and his colleague Antonio Cadeddu confirmed the truth of this version of reality by carefully examining Pasteur's laboratory notebooks. They concluded that Pasteur had "deliberately deceived the public and the scientific community about the nature of the vaccine actually used at Pouilly-le-Fort" (Geison, 1995, p. 151). This was the secret of Pouilly-le-Fort. According to Loir (see Geison, 1995, p. 149): "Pasteur, at that time, pursued the attenuation of viruses by atmospheric oxygen. It was a theory that he had conceived. Oxygen destroyed the virulence of all microbes. The immense role of oxygen was an idea he had long held. He pursued its demonstration." Pasteur vowed that he would not permit anyone to publish the results of his experiments with potassium bichromate until attenuation by oxygen had been found.

Pasteur had held very strong beliefs about the role played by oxygen in destroying harmful disease-causing bacteria. He had concluded earlier that oxygen was the agent that attenuated the virulence of the chicken cholera microbe. So he believed that this same mechanism was probably applicable to the anthrax problem as well. But he had no proof. Nevertheless, he reported to the Academy of Sciences that it was an oxygen-attenuated vaccine that he had used rather than the antiseptic vaccine that he had actually used. He had let his beliefs about which vaccine was appropriate to immunize against anthrax, rather than his actual data, drive his representation to the world.

Pasteur's claims of discovery were in some sense similar to Einstein's many claims (Section 4.13) about the truth of his theories long before there were data that confirmed them. In Pasteur's case, however, the scientist misrepresented the mechanism for the conclusions reached. Moreover, Pasteur did not derive his conclusions by inferences from mathematical formulations, as did Einstein. In both cases, however, these scientists arrived at their conclusions from a profound understanding of the underlying science involved rather than from confirming experiments.

In 1885, Pasteur saved the lives of the boys Joseph Meister and Jean-

Baptiste Jupille, who had been bitten by a rabid dog. Pasteur claimed to have perfected a vaccine against rabies and these cases demonstrated that his vaccine worked. He repeatedly claimed to have previously produced immunity in 50 animals. Yet, on July 6, when he gave the first injection to Meister, Pasteur's notebooks make clear that he had just begun a series of vaguely comparable experiments on 40 dogs. However, he had been trying out various vaccines, and he speculated that his vaccine would work—and fortunately for the boys, it did. On the day that he delivered his famous paper to the French Academy of Sciences, October 26, 1885, Pasteur was still injecting Jupille—not even waiting for the full outcome before claiming success.

We should note that there is some justification for human experimentation of this sort, even in the absence of the animal evidence that Pasteur claimed. The boys were savagely bitten and the chance that they would come down with rabies was very high. If they did contract rabies, death was very likely. Pasteur had a possible cure, and although he had no actual evidence that it would work, he did have his subjective belief that it would. Hence, using it rather than not treating the boys at all may indeed have been ethically justifiable.

According to Geison, Pasteur's work on spontaneous generation was also subjected to some manipulation, for somewhat different reasons. Pasteur was a political conservative who used his science in the service of France, the Catholic Church, and the Second Empire. Pasteur saw to it in these experiments that nature did not speak for itself but was shown to be on the side of the Second Empire, the Catholic Church, and France. Pasteur was careful to exclude from his public announcements any experiments that did not confirm his personal biases. Pasteur labeled results experimental errors if they conflicted with his a priori belief that spontaneous generation must be wrong. Such an idea had to be wrong because it conflicted with his values.

We pointed out in Chapter 1 that the same data may be interpreted differently by different observers. Pasteur understood that data need to be interpreted. Because he was a political animal he interpreted his data in ways that would support the Catholic Church and the French nation. Because he interpreted his data selectively, he misled the public and other scientists as well as the Academy of Sciences. Pasteur had come up with undesirable results in his experiments, so he manipulated the data. He threw out the data that were disturbing. He arranged for his experiments to confirm his judgments.

The following is an adaptation of portions of Geison (1995): We can explain such failures in Pasteur's experiments in the light of the knowledge of today—that some of the organisms Pasteur had in the samples in his experiment were able to survive unless the temperature were extremely high. But in a so-called objective scientific approach, when the data conflict with the scientist's beliefs and theory, the scientist should report the anomalous data, and might change the theory until it conforms to the observed data. Pasteur seems to have sometimes done it the other way around. He apparently ignored the data that conflicted with the strong beliefs he held prior to his taking the data. Pasteur defined as "unsuccessful" any experiments that gave results contrary to his beliefs. Moreover, during the 1860s, the most politically sensitive phase in the spontaneous generation debate, Pasteur's public image was that of fearless crusader against the dangerous doctrine of spontaneous generation. Yet one learns from the notebooks that during those years, Pasteur considered the artificial creation of life a theoretical possibility. He had actually pursued the problem experimentally in the early 1850s. For 30 years, Pasteur said nothing in public about these remarkable experiments and very little about preconceived ideas. The public would have been amazed to learn that Pasteur had managed to wage a vocal public campaign against spontaneous generation even as he speculated about the creations of life in his own special version of this dangerous doctrine (pp. 137–138).

Some conclusions regarding Geison's book, bear repeating. Commenting on the discrepancies between the notebooks and Pasteur's public pronouncements, Higgins (1995) points out that such discrepancies are familiar to working scientists or students of "science studies." For them: "[I]t will be obvious that such discrepancies are part . . . of the process by which 'raw data' are transformed into published 'results.' In the interests of brevity, clarity, logical coherence and rhetorical power, the published record always projects a more or less distorted image of what the scientist 'really' did."

Dubos (1950) knew of Pasteur's preconceptions and some missteps that resulted in other areas as well:

> As will be remembered [Pasteur] had isolated from the saliva of the first rabid child that he studied a virulent encapsulated microorganism—now known as the pneumococcus. Before recognizing that this organism did not bear any relation to rabies, he constructed the theory that the period of incubation of the disease was the time required for the dissolution or destruction in the tissues of the capsule surrounding the microorganism. Had this working hypothesis fitted the facts of rabies, it would have been, indeed, an exciting theory; but because it did not, scientific etiquette rules it bad taste to mention it in print.

> And yet, the elucidation of the mental processes involved in scientific discovery requires a knowledge of the hypotheses that miscarry, as well as of those which bear fruit (p. 376).

We have presented some of the evidence regarding the subjectivity and possible fraud of Louis Pasteur in light of the research of biographers such as Dubos (1950) and Geison (1995). Dubos sometimes chose to overlook apparent inconsistencies in Pasteur's methodological approach to science; Geison used Pasteur's own laboratory notebooks to illuminate seemingly inexplicable misrepresentations and discrepancies between his data and his theories. But Pasteur also had his strong defenders among scientists. Perutz (1995) wrote a long critique of Geison's 1995 book, claiming that Geison had misinterpreted Pasteur's notebooks. But Higgins (December 26, 1995) convincingly countered Perutz, claiming that Perutz was merely trying to uphold Pasteur's sacred place as a hero and an icon in the history of biology.

We have examined the scientific methodology of Louis Pasteur on the basis of the evaluations of his principal biographers, and on the basis of his private laboratory notebooks. We find that Louis Pasteur was a giant among scientists, a pathbreaker in his research, and like other scientists before him, very informally subjective in his methodological approach.

REFERENCES

Altman, Lawrence K. (1995). "Revisionist History Sees Pasteur as Liar Who Stole Rival's Ideas," *New York Times*, May 16, pp. C1, C3.

Collins, Harry, and Trevor Pinch (1993). *The Golem: What Everyone Should Know About Science*. Cambridge: Cambridge University Press.

Dagognet, Francois (1967). *Méthodes et Doctrines dans l'Oeuvre de Pasteur*. Paris: Presses Universitaires.

DeKruif, Paul (1953). *Microbe Hunters*. New York: Harbrace Paperback Library.

Dubos, René (1950, 1960, 1976). *Louis Pasteur: Free Lance of Science*. New York: Scribners, Da Capo Press.

Duclaux, Emile (1895). "Le Laboratoire de Monsieur Pasteur à l'Ecole Normale," *Revue Scientifique*, 4th series, **15**:449–454. Reprinted in *Le Centenaire de l'Ecole Normale, 1795–1895*. Paris: Ecole Normale.

———(1896). *Pasteur: Histoire d'un Esprit*. Sceaux, France: Charaire. English translation by E. F. Smith and F. Hedges as *Pasteur: The History of a Mind* (1920). Philadelphia: W.B. Saunders.

Farley, John, and Gerald Geison (1982). "Science, Politics and Spontaneous Generation in Nineteenth-Century France: The Pasteur–Pouchet Debate," pp. 1–38 in H. M. Collins (Ed.), *Sociology of Scientific Knowledge: A Source Book*. Bath, UK: Bath University Press. [Originally, *Bulletin of the History of Medicine* (1974), **48**:161–198.]

Geison, Gerald A. (1974). "Louis Pasteur," pp. 350–416 in Charles C. Gillispie (Ed.), *The Dictionary of Scientific Biography*, Vol. 10. New York: Scribners.

———(1995). *The Private Science of Louis Pasteur*. Princeton, NJ: Princeton University Press.

Higgins, A. C. (1995). *Science Fraud Bulletin Board* (Internet address: scifraud@albnyvm1), June 12, 13, December 26.

Loir, Adrien (1937–1938). "A l'Ombre de Pasteur" (In the Shadow of Pasteur) *Mouvement Sanitaire*, **14**:91–92.

Medawar, Peter B. (1964). "Is the Scientific Paper Fraudulent," *Saturday Review*, August 1, pp. 42–43.

Metchnikoff, Elie (1933). *Trois Fondateurs de la Médecine Moderne: Pasteur, Lister, Koch*. Paris: Libraire Felix Alcan.

Nicolle, Jacques (1953). *Un Maître de l'Enquête Scientifique, Louis Pasteur*. Paris: La Colombe. Translated as *Louis Pasteur: A Master of Scientific Enquiry*. London: Hutchinson, 1961.

———(1966). *Louis Pasteur: The Story of His Major Discoveries*. New York: Fawcett.

Perutz, M. F. (1995). "The Pioneer Defended," a review of Gerald L. Geison, *The Private Science of Louis Pasteur* (Princeton, NJ: Princeton University Press) *New York Review of Books*, December 21, pp. 54ff.

Roux, Emle (1896). "L'Oeuvre Médicale de Pasteur," in *Agenda du Chimiste*, Paris. Reprinted in *Institut Pasteur*, "Pasteur 1822–1922." Translated by E. F. Smith as "The Medical Work of Pasteur," *Scientific Monthly*, **21**:325–389 (1925).

Vallery-Radot, René (1883). *Pasteur: Histoire d'un Savant par un Ignorant*. Paris: J. Hetzel.

———(1900). *La Vie de Pasteur*, 2 vols. Paris: Flammarion. Translated as *The Life of Pasteur* by R. L. Devonshire, 2 vols. London: Constable, 1911.

Wolpert, Lewis (1995). "Experiments in Deceit: A Review of Gerald L. Geison, *The Private Science of Louis Pasteur*," Book Review, *New York Times*, May 7, p. 35.

4.11   SIGMUND FREUD (1856–1939)

Sigmund Freud was a psychologist who was the founding father of the field we now call *psychoanalysis*.

4.11.A   Brief Biographical Sketch

Sigmund Freud was born in Freiberg, now part of the Czech Republic, in 1856. At that time, Freiberg was in a region called Moravia, ruled by Austria. Although he was Jewish by birth, he considered himself an atheist. Czech nationalism against Austrian rule was on the rise, and the German-speaking Jewish minority offered an easy target for hostile feelings. When Freud was 3 years old, the family moved to Vienna, Austria, to the poor and overcrowded Jewish ghetto of the city. Anti-Semitic incidents were to affect his professional research and belief systems. He lived in Vienna for the next 79 years, almost to the end of his life.

In the educational system of Vienna, to become a medical doctor, one went directly to medical school after secondary school, without an intermediate undergraduate degree. So at the age of 17, Freud entered medical college at the University of Vienna. He was a distinguished medical student and worked in the physiology laboratory of Ernst Brucke. But Freud shunned quantitative (mathematical or physical science) approaches to medical matters in favor of the more qualitative histological and clinical approaches. He obtained his medical diploma at 25.

At 26, Freud abandoned basic science and decided to become practical, so that he could earn a living. Accordingly, to qualify to practice on people, he spent three years doing a residency at Vienna General Hospital. There, among the many other medical specialties he explored, he studied psychiatry under Theodor Meynert, an anatomist of the brain. He joined Meynert's psychiatric clinic as assistant physician in 1882.

At age 29, Freud went to Paris for $4\frac{1}{2}$ months, where he studied under the most famous neurologist of his time, Jean Martin Charcot. This was surely a major turning point in his career, for there that he saw Charcot studying *hysteria*, a condition of the mind. It must have struck young Freud that the study of the psychological aspects of neurology, as Charcot was doing, was far more interesting than the study of the physical aspects of neurology with which he had heretofore been engaged. This was a basic shift in his interests, from the brain to the mind, a shift that was to be permanent. He started private practice as a neurologist at age 30 and married the following year.

The remainder of Freud's life involved his treating patients for mental problems and combining his case studies of those patients with his understanding of biology to try to understand the human psyche by inferring general theories about the causes and possible resolutions of such psychological problems in general. His work resulted in the development of the general field of psychoanalysis. In 1939, the Nazis formally annexed Austria. Freud escaped by fleeing to London. He died there from cancer later that year, at the age of 83.

4.11.B Freud's Scientific Contributions

Freud's contributions to science may be classified into four general categories of topics; his work was carried out during partially overlapping time periods (see Alexander and Selesnick, 1966, p. 241): studies in hypnotism and hysteria (1886–1895), contributions to the anatomy of the nervous system and to neurology (1893–1897), study of the unconscious and the development of the psychoanalytic method of treating patients with problems of the mind (1895–1920), and studies of personality and the structure of society (1920–1939). His work in hypnotism and hysteria resulted in his publication, with Josef Breuer, of *Studies in Hysteria* (in German) in 1895. Freud's research in neurology concentrated on the basic science of the body more than did his studies of the mind, which were based on case studies.

Freud's most productive period, in the sense of his contributions to the development of psychoanalysis, was between 1895 and 1900. In 1895 he developed the method of *free association*, in which he asked patients to abandon conscious control over their ideas and to say whatever came into their minds. Freud reasoned that there are opposite forces at play in the mind: those seeking to repress unpleasant events long since forgotten by the conscious mind, and those seeking to express feelings and emotions—forces of repression and expression. These opposing forces generated conflicts in the mind that resulted in mental and physical disorders. By asking his patients to freely associate, Freud hoped that important thoughts that were previously repressed in the unconscious mind of the patient would come to the surface and be expressed, perhaps inadvertently, by the conscious mind. Once the conscious mind was able to "work through" and therefore to understand these repressed thoughts, the patient's disorders would disappear. Freud referred to this procedure (in 1896) as *psychoanalysis*. Later, others have also applied the term to the process of ridding a patient of his or her neuroses by discussion with the patient, regardless of whether or not the therapist believes in a repression-based genesis of the problem (*non-Freudian psychoanalysis*).

Freud noticed that while undergoing free association, many patients began to talk spontaneously about their dreams, often revealing the hidden content of their dreams. He then began to understand that the dreams were attempts by the patient to relieve emotional tensions and frustrations caused by inner conflicts. It was in 1900 that Freud published what is generally accepted to be his greatest work, *The Interpretation of Dreams*. Over time, Freud became increasing adroit at interpreting his patients' dreams. Moreover, the more he studied the dreams, the more he became convinced that sexual impulses played a significant role in neuroses. In fact, he finally became convinced that sexual conflict is the most influential driving force in most, if not all, psychopathology. His conclusions were published in 1905 in *Three Essays on the Theory of Sexuality*.

It was not until 1920 that Freud finally decided to summarize some of the results of his case studies in terms of a theory of personality, by publishing *Beyond the Pleasure Principle*. He then also began to speculate about societal ills in terms of group psychology, as exemplified by *Civilization and Its Discontents*. These speculations continued until his final work, *Moses and Monotheism*, published in 1939.

As the field of treating mental illness has developed, other researchers and clinicians have introduced alternative explanations of the source of neuroses. They have proposed alternative methods of treatment (such as behavior therapy, psychopharmaceutical therapy, cognitive therapy, and a variety of psychodynamical procedures), and they have suggested experiments that might test Freudian and related theories of the etiology and treatment of mental illnesses.

4.11.C Major Works

Freud, Sigmund (1899). *Die Traumdeutung* (The Interpretation of Dreams). Leipzig: F. Deuticke.

————(1904). *Zur Psychopathologie des Alltagslebens* (Psychopathology of Everyday Life). Authorized English edition (1914) by A. A. Brill. London: T. Fisher Univan.

————(1905). *Drei Abhandlungen zur Sexualtheorie* (Three Essays on the Theory of Sexuality). Leipzig: F. Deuticke.

————(1910). *Uber Psychoanalyse* (The Origin and Development of Psycho-Analysis). Vienna: F. Deuticke.

———(1913). *Totem und Tabu* (Totem and Taboo). English verison, *Totem and Taboo and Other Works* (1955) London: Hogarth Press and the Institute of Psychoanalysis.

———(1914). *Zur Geschuchte der Psychoanalytischen Bewegung* (On the History of the Psychoanalytic Movement). English version (1917), London: The Nervous and Mental Disease Publishing Co.

———(1917). *Vorlesungen zur Einfuhrung in die Psychoanalyse* (A General Introduction to Psychoanalysis). English version (1920), New York: Boni and Liveright.

———(1920). *Jenseits des Lustprinzips* (Beyond the Pleasure Principle). English version (1922), London: The Psycho-analytical Press.

———(1923). *Das Ich und das Es* (The Ego and the Id). English version (1927), London: L. and Virginia Woolf at the Hogarth Press and the Institute of Psychoanalysis.

———(1925). *Selbstdarstellung* (An Autobiographical Study). Translation to English (1935), London: L. and Virginia Woolf at the Hogarth Press and the Institute of Psychoanalysis.

———(1926). *Hemmung, Symptom und Angst* (Inhibitions, Symptoms and Anxiety). Vienna: Intionationaler Psychoanalytischer Verlag.

———(1926). *Die Frage der Laienanalyse* (The Question of Lay-Analysis). English translation (1950) New York: Norton.

———(1927). *Die Zukunft einer Illusion* (The Future of an Illusion). English translation (1928) London: L. and Virginia Woolf at the Hogarth Press and the Institute of Psychoanalysis.

———(1930). *Das Unbehagen in der Kultur* (Civilization and Its Discontents). English translation (1930) London: L. and Virginia Woolf at the Hogarth Press and the Institute of Psychoanalysis.

———(1933). *Neue Folge der Vorlesungen zur Einfuhrung in die Psychoanalyse* (New Introductory Lectures on Psycho-Analysis). Translated by W. J. H. Sprott (1933), New York: W.W. Norton.

———(1939). *Der Mann Moses und die Monotheistische Religion* (Moses and Monotheism). English translation (1985) Harmondsworth: Penguin.

———(1974) *The Standard Edition of the Complete Psychological Works of Sigmund Freud*, 24 vols., James Strachey et al. (Eds.). London: Hogarth Press and the Institute of Psychoanalysis.

Freud, Sigmund, and J. Breuer (1895). *Studien über Hysterie* (Studies in Hysteria). English translation (1950) Boston: Beacon Press.

4.11.D Subjectivity in the Work of Freud

Since Freud's death there have been many who have questioned whether his research methodology was really honest and objective, and even whether it was really science. The assaults on his work have steadily increased. Recently, a variety of documents relating to Freud and his work have been released, some by his estate. These documents have raised many questions, not only about his use of personal beliefs in his scientific research. They go further, to ask also whether the entire field of psychoanalysis he started, and the related field of clinical psychology, should be considered bone fide scientific fields, or just passing pseudoscientific fads (see, e.g., Dawes, 1994). An informative account of the more recent attacks on Freud's work was given by Gray (1993). Not only has the Freudian theory become the dominant view of mental illness and its treatment, but his influence has permeated our culture. We need little familiarity with the actual works of Freud to understand the meaning of such terms as *Freudian slip* or *Oedipus complex*. But the advent and increased success of psychoactive drugs suggest that the etiology of mental illness may be more physiological and less psychological than Freud believed.

Freud and his entire theory of psychoanalysis have come under frontal assault. Adapting from Gray (1993), we can see that some of the accusations were leveled against Freud even quite early. The 10-year collaboration between Freud and Carl Jung ended in 1914, to the detriment of psychoanalysis, when a difficult schism developed in the field that permanently divided its adherents. From that time on there were Freudians and Jungians. John Kerr argues in *A Most Dangerous Method* (quoted in Gray, 1993, p. 49) "... that the growing philosophical disputes between Freud and Jung were exacerbated by a cat-and-mouse game of sexual suspicion and blackmail."

Freud contended that an ex-patient of Jung's named Sabina Spielrein had also been Jung's mistress. On the other side, Jung became convinced that Freud had become involved with his sister-in-law, Minna Bernays. Both men could readily have destroyed each other's reputations. But each of them retreated from these hazardous positions, preferring instead to benefit personally from their collaborative efforts, with the result that their professional positions partitioned different portions of the theory. Gray has asked (1993, p. 49): "Was this any way to found an objective science?"

Freud's supporters have argued that his personal life bears little relation to his contributions to the field, despite the fact that Freud himself stated that his development of the analytic method began with his pioneering analysis of himself. Regardless of this assertion, Arnold Richards, editor of the American Psychoanalytic Association newsletter, dismissed any attention paid to Freud's private conduct: "It has no scientific practical consequence. It's not relevant to Freud's theory or practice" (quoted in Gray, 1993, p. 49). Freud's published case histories largely recorded inconclusive results. For this reason, some of his defenders have argued that he may not have been very good at implementing his own beliefs, but that failing in no way invalidates his theories.

Freud's supporters have also had to defend against the onslaught embodied in Grunbaum (1993). The 1993 book continues Grunbaum's assault (Grunbaum, 1983) on the fundamental structure of psychoanalysis and brings into question whether psychoanalysis should even be considered a science. He accomplishes this by examining a number of key psychoanalytic premises: the theory of repression (which Freud called "the cornerstone on which the whole structure of psychoanalysis rests"), the investigative capabilities offered by free association, and the diagnostic significance of dreams. Although Grunbaum admits that the idea of repressed memories, for instance, may be correct, he questions whether Freud or any of his successors has ever established a cause-and-effect link between a repressed memory and a later neurosis or a retrieved memory and a subsequent cure.

In Freud's favor, Gray (1993, p. 51) cites the work of Jonathan Winson, professor emeritus of neurosciences at Rockefeller University, who has done extensive research on the physiology of sleep and dreams. Winson claims that Freud's intuition of the existence of the unconscious was correct, even if not based on scientific evidence. Winson also credits Freud for sensing that dreams are the "royal road" to the unconscious. In the final analysis, Freud may have gotten so many important concepts right that it is, and will be, difficult simply to brush him aside. We reach this conclusion despite his careless treatment of colleagues and patients alike, and despite his many sins of omission and commission. He still managed to create an intellectual construct that appears to contain more important and basic ideas than does any other psychological system currently extant. But despite his arguments to the contrary, the field of psychoanalysis he established so subjectively has not yet proved itself to be a science.

But Freud's more-than-generous use of subjectivity in his approach to his work did not escape notice even by his contemporaries. For example, Joseph Jastrow wrote (1932, p. 93):

> The initial source of dissension was the far reaching sovereignty assigned by Freud to sex in the psychic life, and particularly to the detailed deductions derived from its imperial sway. The protests against the sexualization of the psyche were many and emphatic, and made not by squeamish Puritans, but by responsible scientists and by Freud's own followers. In Jung's "analysis," sex is as strident as in Freud's; but in his survey the urge to live and live the life abundant cannot be confined to the will to live sexually, nor is it derived from it. Libido includes other urges, biologically parallel and equally primary. Libido is life-energy expressed through the psyche. So radical a reshaping of a major doctrine was proclaimed a heresy. Sex was the shibboleth; those who pronounced it differently were of another tribe. They were met by the excommunication of estrangement.

Jastrow was here noting how strongly Freud held to his personal belief that sexual urge and libido was the end-all, the driving force behind all repression and inner conflict. Freud held fast to this belief, despite major objections and major objectors. It was a very strong *prior belief*, in our technical sense, and one that was not to be modified objectively in the light of observational data, if such could be found.

But what generated such strong prior beliefs by Freud? What were the influences on him that engendered such strongly held personal notions of the drives that must be responsible for molding and shaping human behavior? Torrey (1992, p. 13) states:

[D]uring the 1890s Sigmund Freud was a man with a mystical sense of personal destiny, who believed in telepathy and numerology, and who was using cocaine at least intermittently. Specifically, Freud acknowledged "making frequent use of cocaine" during the period when he analyzed his own dream of Irma's injection in 1895; this dream, labeled by Ernest Jones as "an historic moment," has been called the prototypic "dream specimen" of psychoanalysis. The dream of Irma's injection also inaugurated Freud's own self-analysis, during which he developed his theory of childhood sexual development and the Oedipal complex.

The fact that Freudian theory evolved simultaneously with a sense of destiny, an interest in the occult, and the use of cocaine does not in itself negate the validity of the theory. It does, however, cast a shadow over its scientific foundation. It also makes more comprehensible Freud's own recollections of the development of his theory, published in 1914 as a pamphlet titled *The History of the Psychoanalytic Movement*: "Since, however, my conviction of the general accuracy of my observations and conclusions grew and grew, and as my confidence in my own judgment was by no means slight, any more than my moral courage, there could be no doubt about the outcome of the situation."

Torrey (1992, pp. 215–216) also wrote: "In later years, when a psychologist wrote Freud to tell him that he had found scientific support for Freud's theory, Freud responded testily that his theory needed no validation." This statement of Freud clearly demonstrates his very strong personal belief in his own theory, regardless of data, a characteristic shared by Newton, Einstein, Pasteur, and many of the other scientists being studied in this book. This informal subjectivity in his methodological approach to research is parallel to the informal subjectivity employed by the other famous scientists we are examining in this chapter, and contrasts with the Bayesian approach of using formal subjectivity mixed with experimental data advocated in Chapter 5.

Torrey (1992) continued with a statement of Freud in the same vein: "It should also be recalled that Freud was actively conducting experiments on occult phenomena throughout these years, and it may have been within such a context that Freud spoke when he once proclaimed: 'We possess the truth, I am sure of it.'" Kline (1972) added: "Freudian theory, so far as it is dependent on data at all, rests on data which by the criteria of scientific methodology are totally inadequate. These data are, for the most part, the free associations of patients undergoing therapy and their dream reports, and both of these sources are unquantifiable and riddled with subjective interpretation."

In a review of Freud's work as "science" (Eysenck and Wilson, 1973), the authors concluded: "In this volume the authors found not one study which one could point to with confidence and say, 'Here is definitive support of this or that Freudian notion; a support which is not susceptible to alternative interpretation, which has been replicated, which is based on a proper experimental design, which has been submitted to proper statistical treatment, and which can confidently be generalized, being based on an appropriate sample of the population.' After three quarters of a century, this is a serious indictment of psychoanalysis."

Another important influence on Freud's development of psychoanalysis and the building of his very strong prior beliefs was his ethnic background (see, e.g., Bakan, 1958; Grollman, 1965). Grollman wrote:

Before the Freudian era, a movement was led by a man who represented the complete opposite of cold pedantry. The new approach followed the uprisings in the Ukraine when more than 100,000 Jews lost their lives in less than a decade. [*Note from the authors*: Israel ben Eliezer, generally known by the name of Israel Baal Shem Tov (1700–1760), about 1735, started a movement called Chassidism (Hebrew, meaning "piousness") that spiritually revived Slavic Jewry and deeply penetrated all Jewry in Eastern Europe.] In his movement of Chassidism, a man could literally escape his unbearable miseries by immersing himself in a mystical *hithlahabuth* (the esoteric kindling of the soul with God). To the masses who hungered for a direct, simple, stimulating religion which they could follow without any philosophical sophistications, the doctrine of salvation through prayer and humility rather than study was appealing. The unsupressed emotions and optimistic Chassidic spirit served as a buffer against the depressing environment of dissolution and terror.

It is important to note that many of the pioneers in psychoanalysis were born of parents upon whom Chassidism had made an impact. Freud acknowledged his father Jakob came from Chaissidic stock, even as Josef Breuer's father was a Chaissidic rabbi. Professor David Bakan postulated that Freud gave birth to psychoanalysis because his mind was already pregnant with Chassidic mysticism. He mentioned Freud's awareness of the 16th century Jewish mystical physician, Solomon Almoli, whose book, *The Solution of Dreams*, gives a description of sexual symbolism, wish fulfillment and word play as elements found in dreams. Many counterparts of Freudian theory were found in the Zohar (literally "brightness," the mystical writings issued in part by Moses de Leon in the 14th century) such as the portrayal of primordial man where the divine act of creation was given an erotic character and where sex relations were treated as avenues of salvation.... E. Steindletz also mentioned the prior belief of Chassidism in an unconscious sex drive and pleasure principle.

Bakan states in the preface to his 1958 book: "Freud's repeated affirmation of his Jewish identity had greater significance for the development of psychoanalysis than is usually recognized. He was a participant in the struggles and the issues of Jewish mysticism; and where it was appropriate, he drew from the Jewish mystical armamentarium for equipment in these struggles."

Shweder (1995) pointed out how different disciplines have different objections to various parts of Freud's subjectively derived system in a commentary on a decision of the Library of Congress to cancel a planned exhibition of Freud's work. He concludes by saying (p. 19): "And the philosophers of science come in two kinds: those who think Freud's tenets are untestable and hence unworthy of scientific consideration, and those who think his tenets are testable and have been shown to be false."

In summary, in the case of Sigmund Freud, it is not necessary to search deeply for specific instances in which subjectivity played a role in his research methodology; his subjectivity is very visible throughout his work. Subjectivity was an integral part of Freud's methodology and clearly dominated his conclusions throughout his development of psychoanalysis. Freud's contributions to science are under attack at the present time, as were the works of Darwin and Galileo during their lives and after their deaths. It will therefore only be in the light of historical perspective that we will be able to assess whether Freud's use of personal beliefs, biases, and intuition were helpful to his contribution of new and important results in science, or misleading toward error.

REFERENCES

Abraham, Hilda C., and Ernst L. Freud (Eds.) (1965). *A Psycho-Analytic Dialogue: The Letters of Sigmund Freud and Karl Abraham, 1907–1926.* London: Hogarth Press.

Alexander, F., and Sheldon Selesnick (1966). *The History of Psychiatry.* London: Hogarth Press.

Andreas-Salome, Lou (1965). *The Freud Journal of Lou Andreas-Salome* (English translation of *In der Schule bei Freud: Tage-buch eines Jahres, 1912–1913*). London: Hogarth Press.

Bakan, David (1958a). "Moses in the Thought of Freud," *Commentary,* October, pp. 322–331.

———(1958b). *Sigmund Freud and the Jewish Mystical Tradition.* Princeton, NJ: D. Van Nostrand.

Bonaparte, Marie, Anna Freud, and Ernst Kris (Eds.) (1887–1902). *Sigmund Freud, the Origins of Psychoanalysis: Letters to Wilhelm Fliess, Drafts and Notes,* English translation by Eric Mosbacher and James Strachey. New York: Basic Books.

Bottome, Phyllis, and Alfred Adler (1946). *Apostle of Freedom,* 2nd ed. London: Faber & Faber.

Brome, Vincent (1967). *Freud and His Early Circle.* New York: Morrow.

Costigan, Giovanni (1968). *Sigmund Freud: A Short Biography.* New York: Collier Books.

Dawes, Robyn (1994). *House of Cards: Psychology and Psychotherapy Built on Myth.* New York: Free Press.

Encyclopaedia Britannica Macropedia (1975). Vol. 7, p. 742.

Eysenck, Hans, and Glenn D. Wilson (1973). *The Experimental Study of Freudian Theories.* London: Methuen.

Fliess, Wilhelm (1906). *In Eigener Sache: Gegen Otto Weininge und Hermann Swoboda.* Berlin: E. Goldschmidt.

Freud, Ernst L. (Ed.) (1960). *Letters of Sigmund Freud, 1873–1939,* translated by Tania and James Stern. New York: Basic Books.

Fromm, Erich (1959). *Sigmund Freud's Mission: An Analysis of His Personality and Influence.* New York: Harper.

Gray, Paul (1993). "The Assault on Freud," *Time,* November 29, pp. 47–51.

Grollman, Earl A. (1965). *Judaism in Sigmund Freud's World*. New York: Bloch Publishing Co.

Grunbaum, Adolf (1983). In R. S. Cohen and L. Laudan (Eds.), *Physics, Philosophy, and Psychoanalysis: Essays in Honor of Adolf Grünbaum*. Dordrecht, The Netherlands: D. Reidel (distributed in the United States by Kluwer Boston).

———(1993). *Validation in the Clinical Theory of Psychoanalyis: A Study in the Philosophy of Psychoanalysis*. Madison, CT: International Universities Press.

Jaffe, Aniela (Ed.) (1963). *Carl Jung: Memories, Dreams, Reflections*, translated from the German. New York: Vintage Books.

Jastrow, Joseph (1932). *Freud: His Dream and Sex Theories*. New York: Pocket Books.

Jones, Ernest (1953–1957). *The Life and Works of Sigmund Freud*, 3 vols. New York: Basic Books.

Kline, Paul (1972). *Fact and Fancy in Freudian Theory*. London: Methuen.

Peters, Heinz E. (1962). *My Sister, My Spouse: A Biography of Lou Andreas-Salome*. New York: W.W. Norton.

Puner, Helen W. (1947). *Freud: His Life and His Mind, A Biography*. New York: Howell, Soskin.

Reik, Theodor (1940). *From Thirty Years with Freud*. London: Hogarth Press.

Robert, Marthe (1964). *La Revolution Psychoanalytique: La Vie et l'Oeuvre de Sigmund Freud*, 2 vols.; English translation, *The Psychoanalytic Revolution: Sigmund Freud's Life and Achievement* (1966). New York: Harcourt, Brace and World.

Sachs, Hanns (1944). *Freud: Master and Friend*. Cambridge, MA: Harvard University Press.

Shweder, Richard A. (1995). "It's Time to Reinvent Freud," Op Ed., *New York Times*, December 15, p. A19.

Stekel, Wilhelm (1950). *Autobiography: The Life Story of a Pioneer Psychoanalyst*. New York: Liveright.

Torrey, E. Fuller (1992). *Freudian Fraud: The Malignant Effect of Freud's Theory on American Thought and Culture*. New York: HarperCollins.

Trilling, Lionel, and Steven Marcus (1961). *The Life and Works of Sigmund Freud*, edited and abridged. New York: Penguin Books.

Wittels, Fritz (1924). *Sigmund Freud*, English translation. Leipzig: E.P. Tal.

Wortis, Joseph (1954). *Fragments of an Analysis with Freud*. New York: Simon & Schuster.

4.12 MARIE SKLODOVSKA CURIE (1867–1934)

Marie Curie won Nobel prizes in both physics and chemistry for her discoveries of the elements polonium and radium.

4.12.A Brief Biographical Sketch

Marie Curie was born Marya Sklodovska in Warsaw, Russian-occupied Poland, on November 7, 1867, the youngest of five children. Because both her parents were teachers, Marya was constantly exposed to teaching and learning. She was extremely precocious, learning to read fluently by age 4 and early showing evidence of an extraordinary memory. Marya graduated from secondary school with a gold medal in 1883 and at her father's insistence took a year of vacation to decide on a future course for her life. On her return to Warsaw she began giving private lessons to supplement her family's income. Both Marya and her older sister Bronya were eager for further study, but women were denied access to the University of Warsaw. The sisters became participants in the "Floating University." This arrangement consisted of lessons on such topics as anatomy, natural history, and sociology, given in secret in private homes through the benevolence of teachers willing to risk imprisonment for the sake of imparting knowledge to young people eager and brave enough to run similar risks for the sake of learning. To help her older sister study medicine in Paris and to earn money for her own study there, Marya worked as a governess for three years. During that time, late into the night she read texts of literature, sociology, and science and worked mathematical problems sent to her by her father, thus acquiring the habit of independent work that would serve her so well in the future. Thereafter, as her father had secured a pension, she returned to Warsaw to live with him. Here, to earn her fare to Paris, she gave private lessons, and for the first time had access to a laboratory in which she tried out some of the experiments described in the chemistry and physics texts she had been reading. In the fall of 1891, at age 24, Marya finally left for Paris to begin her advanced studies.

To start, Marya lived with her sister and brother-in-law, by then both physicians, and registered for courses in physics at the Sorbonne in the Faculty of Science as Marie Sklodovska—the French form of her first name would be used thenceforth by everyone but her family. She found immediately that her command of French was not as perfect as she had thought, and her self-taught knowledge of physics left her well behind her classmates, who were holders of the French baccalaureate. She studied hard to remedy these deficiencies, and delighted in the opportunity to study science at last. Later, to secure more privacy to study, Marie decided to move into student lodgings in an attic in the Latin Quarter. She was able to afford only scant furnishing, a small amount of coal to ward off the winter's chill, and very modest amounts of food, but her time was her own to devote to her studies. She ate so little that she became ill of starvation; when her brother-in-law and sister found out, they took her again into their home to get some food into her. But as soon as she had regained her strength, Marie returned to her attic and her extremely modest diet. In 1893 she passed first in her class on the examination for a master's degree in physics (Licence ès Physiques) and in 1894 second on that for mathematics (Licence ès Mathématiques).

At the beginning of 1894, Marie had received a commission to study the magnetic properties of various steels from the Society for the Encouragement of National Industry. She had no workspace in which to accomplish the task. She confided her problem to M. Kovalski of the University of Fribourg, who was visiting in Paris. Professor Kovalski suggested that the scientist Pierre Curie might have some advice and arranged a meeting between him and Marie. By this time, Pierre had already investigated the phenomenon of piezoelectricity and invented, with his brother Jacques, a device that uses the effect in quartz crystal. This device later would be useful to Marie in her research on radioactivity. He served as chief of the laboratory at the School of Physics and Chemistry of the City of Paris, a post neither prestigious nor very remunerative. The two took a liking to one another immediately, spending the evening discussing matters scientific. They saw each other during the remainder of the school year and remained in correspondence during the summer when Marie returned to Poland. By this time Pierre had suggested marriage, but Marie was reluctant to think about leaving Poland permanently. It took another year of persuasion, but on July 26, 1895 the couple was married. They spent their honeymoon bicycling through the countryside, a pastime that was to be their recreation through the years.

Marie settled in to doing the research on the magnetism of steels in the laboratory and in learning how to keep house. She passed first in the examination for a fellowship in secondary education. On September 12, 1897, Marie gave birth to her first child, Irène, and during this period also published her first scientific monograph, on the magnetization of tempered steel. She then began to think about her doctoral thesis and settled on an investigation of "uranic rays," which had been discovered by Antoine Henri Becquerel in 1895. Within a year Marie (together with Pierre) had made several startling discoveries. They found that the intensity of the radiation released by uranium was proportional to the quantity of the element present and hence was an atomic property. They also found that thorium too had this property, which Marie called *radioactivity*. Thus there were two new radioactive elements, named by the Curies *polonium* and *radium*. During the next four years, as Pierre worked to determine the properties of radium, Marie devoted herself to accumulating a sufficient quantity of radium to establish its atomic weight. The radium was extracted by crystallization from enormous quantities, literally tons, of pitchblende. The Curies obtained the pitchblende as industrial waste, after the uranium had been extracted. After years of heavy work under very poor conditions—a leaky and barely heated shed served as laboratory—Marie extracted 1 decigram of radium in 1902. During this time, to earn enough money to make ends meet, both Curies carried heavy teaching loads (Pierre at the Physics, Chemistry, and Natural Science Annex of the Sorbonne and Marie at the Higher Normal School for Girls at Sèvres, near Versailles). It would have been far easier for them if Pierre had been able to gain the professorship in physical chemistry at the Sorbonne for which he had applied in 1898. But he was rejected, primarily because he had not gone through either the normal schools or the polytechnic, and hence did not have the support such institutions offered their graduates. Pierre had been offered a chair in physics in Geneva in 1900, together with a laboratory, a comfortable salary, and a post for Marie, but the couple felt that the move to Geneva would interrupt their work on radioactivity unduly and so declined the offer. Pierre was proposed for membership in the Académie des Sciences in 1902, but his candidacy failed. Both members of the couple were in poor health from overwork and lack of proper nutrition; Marie suffered a miscarriage in 1903. But also in 1903, Marie received her doctorate for the work with radium and shared the Nobel Prize in Physics for the study of uranium and other radioactive substances with Pierre and Antoine Henri Becquerel.

Other honors also started coming to the Curies; Pierre was invited to lecture at the Royal Institution in London, and the couple was awarded its Davy Medal; Marie received the Osiris Prize. The couple meanwhile decided not to take out a patent for the process for extracting radium and therefore never profited from that development. Hence the prize money from the Nobel Foundation was most welcome; it permitted Pierre to give up his teaching burden and also permitted the hiring of a laboratory assistant. At the end of 1904, the Curies' second daughter, Eve, was born. Finally, in 1904 a chair in physics at the Sorbonne was created for Pierre; the appointment included a laboratory, of which Marie was officially appointed chief. In 1905, Pierre was finally elected to the Académie des Sciences.

The couple continued to dedicate themselves to their laboratory, living quietly and modestly and working long hours. On April 19, 1906, Pierre was run over by a horse-drawn wagon in Paris and died instantly. Marie was hard hit by the loss and took a good deal of time to recover; indeed, she was to be characterized by her daughter Eve as thereafter not only a widow but at the same time a pitiful and incurably lonely woman. Nevertheless, within a month she was back in the laboratory, mindful of what Pierre had said to her some years earlier when she wondered how one of them could go on if the other should die: "Whatever happens, even if one has to go on like a body without a soul, one must work just the same . . ." (cited by E. Curie, 1939, p. 254). She was given the chair in physics at the Sorbonne that had been created for Pierre, thus becoming the first woman to hold such a position in French higher education. This was a lectureship; in 1908 she was given the title of professor.

An Institute of Radium was established; Marie became director of the half of the institute that dealt with the pure research on the physics and chemistry of radioactivity; the other half concentrated on research on the application of radioactivity to biological problems and the treatment of disease. Marie continued her research with thoroughness, precise elucidation of details, and exact measurement. She supervised many students and demanded and received careful and excellent work from them. In 1910 she published an encyclopedic treatment of her work and that of others, entitled *Radioactivity*, with André Debierne isolated metallic radium by electrolysis, and provided a gram of pure radium chloride which in 1911 established the International Radium Standard.

Although 1911 brought failure of election to the Académie des Sciences, it also brought a second Nobel prize, this one in chemistry awarded to her alone, for her work in isolating radium. It also brought a scandal, in which what seems to have been an illicit affair between Marie and Paul Langevin, a colleague of hers and a former student of Pierre's, was discovered by Langevin's wife and widely written about in the tabloid press of the day. A duel was held, a trial was scheduled, and the Swedish Nobel committee suggested that Marie refrain from coming to Stockholm to receive the prize until after the trial. She refused to postpone her trip to Sweden, arguing that there was no connection between her scientific work and the facts of her private life. But the stain of scandal clung to her and to her daughters, her health suffered, and she was never again completely well.

During World War I Marie organized a corps of vehicles each equipped with a portable x-ray apparatus to visit field hospitals and x-ray the wounded. She drove one such vehicle herself, and she and her daughter Irène were active in training technicians to take and read the x-rays. By 1920 her eyesight started deteriorating badly because of cataracts, and her general health declined due to anemia, which resulted from her long years of working with radioactive substances without knowledge of their damaging effects and thus without protection. She traveled twice to the United States, where she was greeted with acclimation and presented by President Harding with a gram of radium to be used in her research and bought with funds raised by popular subscription. She also visited Warsaw, where a great Radium Institute had been opened. She was elected a vice-president of the League of Nations' International Committee on Intellectual Co-operation and worked with the Committee on problems of international scientific bibliographies, agreements on scientific symbols, and the preparation of tables of scientific constants.

It seems that Marie's attachment to radioactivity helped obscure for her the dangers it posed to human health. Of course, these dangers were hard to see: first, because the length of time before the onset of symptoms varied but was usually rather long, and second, because there is a good deal of individual variation in susceptibility to the hazards of radiation. Although many in her laboratories suffered ill effects and several died, her own long-term exposure, which eventually probably caused her death, was very slow to have its worst effects. Nevertheless, on July 4, 1934, she died of pernicious anemia. She was buried beside Pierre in the family grave at Sceaux.

4.12.B   Curie's Scientific Contributions

Marie Curie's major scientific contributions were the discovery of polonium and radium and the investigation of radioactivity. Let us look at these achievements in more detail.

X-rays had been discovered in 1895 by Wilhelm Röntgen in Germany and had become an important subject for laboratory research and a novelty in popular culture. Discharging an electrical current through the gas remaining in a near-vacuum tube would produce a ray that would penetrate many opaque substances and could expose a photographic plate. Thus, x-rays could be used to view the bones of human beings through their flesh and to locate fractures. Shortly thereafter, Antoine Henri Becquerel at the Paris Polytechnic started exploring whether fluorescent bodies under the action of light emitted rays similar to x-rays. In particular, Becquerel was working with salts of uranium and found that such salts emitted rays of unknown nature without exposure to light and that they rendered the surrounding air conductive of electricity. He had observed what Marie Curie was later to dub *radioactivity*, and published several papers on the phenomenon. But having begun by studying phosphorescence, Becquerel considered the emission produced by radium not as something entirely new but only as the first example of a metal exhibiting an invisible phosphorescence.

Becquerel could arouse very little interest among other scientists in his so-called "uranic ray," although William Thomson, Lord Kelvin, had, in 1897, presented a paper before the Royal Society of Edinburgh which confirmed that uranic rays did electrify the air. Lord Kelvin had earlier been interested in the electrometer developed by Pierre and Jacques Curie, had praised it extravagantly, and had made it a point to call on Pierre in his laboratory on a visit to Paris in 1893. Lord Kelvin used the instrument developed by the Curie brothers to make the measurements reported in 1897. It was perhaps the connection between Lord Kelvin and the Curies that helped spark the couple's fascination with uranic rays. In any case, Marie chose the subject as the topic of her doctoral dissertation.

The first step was to measure the amount of what we now call radiation given off by various substances to see if the property found for uranium was peculiar to that element. The Curies invented and built an ionization chamber in which an electric charge was applied to a plate coated with the substance being tested. Whether and how fast that charge was transmitted to another plate separated from the first by 3 centimeters was an indication of the amount of radiation being produced by the substance being tested. The experimental setup used both the electrometer and the very sensitive piezoelectric quartz balance that had earlier been developed by the brothers Curie. Testing uranium, Marie found that the amount of radiation did not depend on either the physical or the chemical state of the element, only on the amount present. This established that radioactivity, as she dubbed this property, was an atomic property. She also tested various other elements and on February 17, 1898, tested the compound pitchblende, which contains uranium but which produced a much stronger current than that produced by uranium alone. A week later she found that thorium was also more active than uranium, although less so than pitchblende. Within a month the Curies (Pierre had dropped his work on crystals to cooperate with Marie in this new endeavor) had developed the hypothesis that pitchblende contained a new element that was more radioactive than uranium. It was difficult to confirm that hypothesis using pitchblende, as that mineral contains many elements and is thus difficult to simulate in the laboratory. But when the Curies discovered that another, much simpler mineral, chalcite, was also radioactive, they were able to simulate it in the laboratory by combining uranium and copper phosphate. The laboratory-produced mineral showed no greater radioactivity than would be expected from the uranium it contained. This finding demonstrated that the naturally occurring chalcite must contain a powerfully radioactive substance other than uranium. On April 12, 1898, Marie established her priority for these discoveries by arranging that a note be read at the Académie des Sciences noting that pitchblende and chalcite exhibited more radioactivity than uranium and that such radioactivity implied the presence of an unknown element. This paper, at least implicitly, represented two other major breakthroughs: the use of radioactive properties to identify new substances and the notion that radioactivity is an atomic property.

The Curies next attempted to isolate the substance in pitchblende that was so radioactive. With the help of the chemist Gustave Bémont they developed a heating treatment that concentrated the active substance. They soon found that two very active substances seemed to result from the breakdown of pitchblende, one accompanying bismuth and one accompanying barium. They worked with the former and felt, although they could not separate it from the bismuth in order to identify it clearly through spectroscopy, that they had sufficient evidence, in particular its radioactivity (400 times that of uranium), that they were dealing with a new element. They named their discovery *polonium* after Marie's native country and wrote a report that was read by Becquerel to the Académie des Sciences on July 18, 1898, announcing their discovery. (It was in the title of this paper, "On a new radio-active substance contained in pitchblende," that the term *radioactivity* was introduced.)

There is a gap of several months in the laboratory notebooks at this point—the Curies were on summer vacation. When they returned they worked with the substance that accompanied barium in the breakdown of uranium; repeated heat concentrations led to a substance that had 900 times the radioactivity of uranium; now spectroscopy yielded a pattern that was attributable to no known element. The Curies tried briefly to isolate the active substance, which they had named *radium*, in order to determine its atomic weight and its actual radioactivity, but failed for the time being. At the end of December 1898, they nevertheless announced their new discovery in a paper coauthored with Gustave Bémont and read before the Académie des Sciences, "On a new strongly radio-active substance contained in pitchblende."

Now it was necessary to isolate the pure elements and measure their atomic weights, thus giving them places in the periodic table and firmly proving that they were indeed new elements. It proved easier to isolate radium from the barium chloride with which it was associated than to isolate polonium, so Marie started with radium. Guessing that radium constituted only 1 part in perhaps 100 of pitchblende, the Curies realized that they would need enormous quantities of the mineral. It turned out, eventually, that radium constituted only 1 part in a million of pitchblende, and that the extraction of radium from a ton of pitchblende required 50 tons of water and 5 to 6 tons of other chemicals to yield 5 to 6 grains of radium. The Curies were able to obtain large quantities of pitchblende as industrial waste after the commercial extraction of uranium, and a large shed, which, although far from comfortable, permitted Marie to undertake the gigantic task of extraction. She developed methods of crystallization that required long, hard, and heavy labor over four years, but she finally succeeded, and these were the methods that were later used commercially as well.

Another strand in the research was seeking an understanding of the nature of the phenomenon of radioactivity. One can start tracing this development with the work of J. J. Thompson at the Cavendish Laboratory at Cambridge University, who used Röntgen's vacuum tube to study the streams of light that run from the negative to positive anodes and which he called *cathode rays*. He concluded that cathode rays were made up of negatively charged particles, much smaller than atoms (which came to be called *electrons*), and he considered these particles the basic building blocks of all matter. This understanding presupposed an "atomist" view of the nature of matter, one that accepted the notion of discrete, although invisible molecules and atoms rather than a more accepted view that postulated a continuous ether holding everything together. But Thompson was going even a step further and postulating that the atom, far from being the immutable and indivisible ultimate building block of nature as Mendeleev had argued, was itself made up of smaller particles. Ernest Rutherford, then at McGill University but earlier a student of Thompson's, studied the emissions from radioactive substances and in 1899 he found that they were made up of at least two kinds of rays: *alpha rays*, which carried a large positive charge but were deflected by flimsy barriers, and *beta rays*, which were negatively charged but penetrated thick barriers. Within the year the Curies (with Becquerel) published a series of papers which showed that beta rays (a component of radioactivity) were identical to electrons; it was later shown that the alpha rays are identical to helium nuclei.

Next, Rutherford showed in a paper published in January 1900, that an "emanation" given off by thorium was different from the parent element and could induce long-lasting radioactivity in any substance on which it fell. But the Curies had published a paper entitled "On the radioactivity induced by Becquerel rays" on November 6, 1899 showing the occurrence of the phenomenon. They attributed the induced radioactivity to a transfer of energy, a kind of phosphorescence. Rutherford's next paper, published in February 1900, noted the Curies' results but dismissed phosphorescence as an explanation. So the question and its implications became, as Quinn (1995, p. 167) explains: "Was the temporary radioactivity, which both Rutherford and the Curies had observed in objects placed in proximity to radioactive substances, caused by a transfer of energy of some kind from the substances themselves, or was it caused by a breakdown product, an offspring which differed in makeup and in behavior from its parent? If another radioactive substance could be found to produce an 'emanation' different from itself, as thorium did, then perhaps such emanations were a general phenomenon of radioactivity." And perhaps elements were transmutable as they transferred their radioactivity.

Indeed, results seemed to show that there were other transformations. An English chemist, Sir William Crookes, treated uranyl nitrate chemically and arrived at a substance that was chemically different from uranium and highly radioactive. He called it *uranium X*. In 1900, André Debierne found a new element, named it *actinium*, and suggested that it might be the same as the thorium emanation. In addition, Debierne found that the barium that had precipitated from a solution of actinium and barium had become radioactive. The Curies themselves noted that everything in their lab became radioactive and attributed this phenomenon to the continual formation of radioactive gas. Rutherford and a colleague then published a paper claiming that thorium emanation, now called *thorium X*, was a different substance from thorium—and when it was produced the parent thorium temporarily lost its radioactivity. The Curies rejected this transformation theory, arguing that lost radioactivity was always regained over time. They still believed that either the radioactivity released was already in the element in the form of potential energy or that radioactive elements draw the energy they release from outside themselves. But Rutherford and his colleagues, by 1903, had concluded that transformation, subatomic chemical change, was the explanation of radioactivity, thus setting the stage for the development of nuclear physics.

4.12.C Major Works

Most of Marie Curie's major works have appeared in various translations. No systematic attempt has been made to list those.

Curie, Marie (1904). *Radio-active Substances*, 2nd ed. London: Chemical News Office (doctoral dissertation.) (A translation, published by the Philosophical Library, New York, was published in 1961.)

————(1921). *La Radiologie et la Guerre*. Paris: Librarie Felix Alcan.

————(1925). *L'Isotopie et les Elements Isotopes*. Paris: Edite par la Societe Journal de Physique.

————(1932). *Les Rayons α, β, y des Corps Radio-actifs en Relation avec la Structure Nucleaire*. Paris: Hermann.

————(1935). *Traité de Radioactivité*. Paris: Hermann. (Originally published in 1910 in Paris by Gauthier-Villars, includes her doctoral thesis, her Sorbonne class lectures, and everything known about radioactivity to that date.)

————(1954). *Oeuvres, Recueillies par Irène Joliot Curie*. Warsaw: Pánstwowe Wydawn. Naukowe.

4.12.D Subjectivity in the Work of Curie

Perhaps the most obvious example of subjectivity in the work of Marie Curie was her insistence, in the face of doubt, that radium was indeed a previously unknown element. Eve Curie in her intimate biography of her mother quotes her (1937, p. 157) talking to her sister: "You know, Bronya, the radiation I couldn't explain comes from a new chemical element. The element is there and I've got to find it. We are sure! The physicists we have spoken to believe we have made an error in experiment and advise us to be careful. But I am convinced that I am not mistaken." Marie Curie was to carry out the backbreaking work of isolating a sufficient quantity of radium to prove that the substance was an element and that she was not mistaken.

A 1902 paper entitled "Sur les corps radio-actif" offered the Curies' own reservations about the role of subjectivity in science, particularly the care that must be taken with hypothesis formulation. Quinn (1995, p. 169) gives a quotation in English:

In studying unknown phenomena we can make quite general hypotheses and advance step by step in concordance with experience. This sure, methodical progress is necessarily slow. In contrast, we can make bold hypotheses in which the mechanism of phenomena is specified. This procedure has the advantage of suggesting certain experiments and above all of facilitating the thought process, making it less abstract by the use of an image. On the other hand, we cannot hope to imagine a priori a complex theory which agrees with experience. Precise hypotheses almost certainly contain a portion of error along with a portion of truth.

Surely this stance made it difficult for the Curies to accept Rutherford's transformation theory until the evidence was overwhelming; and surely this "slow and steady" philosophy characterized Marie Curie's approach to science.

An example of how Marie's research goals influenced her findings is given by her interpretation of an increase in radiation found in her laboratory. In 1898 Marie and Gerhard Carl Schmidt independently discovered that thorium is radioactive. In the course of measuring the ionizing power of thorium prior to making the announcement of the discovery, Marie noticed that there was a slight buildup of activity over time. Thinking it experimental error, she repeated her observations but found the same result. Because her goal at the time was to survey many substances in order to discover which were radioactive, and because she had no preconceived idea about the buildup of radioactivity, she ignored the slightly anomalous findings about the amount of radioactivity associated with thorium. Thus she missed discovering the "emanations" later seized upon by Rutherford and used by him as the basis for understanding the nature of radioactivity and for elaborating a theory of matter.

Of great importance was the preconceived belief in the atomist nature of matter that both Curies held, important not only to their search for the mechanism of radioactivity but also to their ability to identify polonium and radium as elements. Had they subscribed to the notion of an ether, as was not uncommon among scientists of the time, they would not have been able to use the fact that the amount of radiation given off by a substance was proportional to the amount of the radioactive substance present to aid in the discovery of the source of the radioactivity. On the other hand, their belief in the scientific principles they held, in particular the first law of thermodynamics, that energy can be converted from one form to another but in a closed system cannot be created or destroyed, kept them from recognizing that energy is released in radioactive decay. Their preconceptions, together with the fact that they were working with elements with very long half-lives, thus kept them from making the leap to understanding how radioactivity was linked to the atomic makeup of the substances they were studying. Because their preconceived beliefs kept them from accepting the notion of transmutation of elements, they offered two hypotheses, both of which turned out to be false, to explain the origin of the energy given off by radioactive substances. Either the law of conservation of energy was suspended for these special substances, allowing them to create energy de novo, or the energy came from outside. Indeed, Marie speculated that all space may be full of rays that can only be absorbed by elements with high atomic weights.

An interesting aspect of the Curie's subjectivity was why it was difficult for them, especially Marie, to understand radiation as the release of subatomic particles. They had from the beginning considered radioactivity an atomic property, and Marie toyed with the idea of some kind of transformation. Quinn summarizes an argument she made in an article in the *Revue Scientifique* in July 1900:

> Radioactive substances . . . may be "substances in which there is a violent interior movement of substances in the course of breaking up. If this is the case, radium should constantly lose weight. But the smallness of the particle is such that, although the electric charge sent out into the atmosphere is easy to detect, the corresponding mass is absolutely insignificant . . . it would take millions of years for radium to lose . . . milligrams of its weight." Thus it is impossible to measure the loss. "The materialist theory of radioactivity is very seductive," she continues. "It explains the phenomena of radioactivity well."

But the Curies were right in maintaining that it would be impossible to measure the loss of weight—at least it was impossible for them. Further, they had spent many years of their professional lives establishing that radium and polonium were indeed elements. If they then had to accept that these new elements could change over time (presumably into other elements), such acceptance might well threaten their conclusions about the elementary nature of these substances. Thus, as Quinn puts it, theorizing that elements transform into one another (1995, p. 172) ". . . would not only contradict the inviolability of the atom, it would also threaten the inviolability of their new elements." Thus, Marie Curie's subjectivity was embodied in the enormous persistence and tenacity of her scientific beliefs, both those that were correct and those that proved mistaken. Or as Einstein put it in a letter to her in December 1923, she was a person "full of goodness and obstinacy at the same time. . . ."

REFERENCES

Crowther, J. G. (1995). *Six Great Scientists: Copernicus, Galileo, Newton, Darwin, Marie Curie, Einstein*. New York: Barnes & Noble.

Curie, Eve (1939). *Madame Curie: A Biography by Eve Curie*, translated by Vincent Sheean. New York: Doubleday, Doran.

Giroud, Françoise (1986). *Marie Curie: A Life*, translated by Lydia Davis from *Une Femme Honorable*, copyright Librarie Arthème Fayard, 1981. New York: Holmes & Meier.

Pflaum, Rosalynd (1989). *Grand Obsession*. New York: Doubleday.

Quinn, Susan (1995). *Marie Curie*. New York: Simon & Schuster.

4.13   ALBERT EINSTEIN (1879–1955)

Albert Einstein was a world-renowned physicist whose numerous far-reaching scientific contributions have affected many aspects of the modern world. He was considered by many in his own time to be the greatest scientific genius the world has ever known. For many people, he is most closely associated with the theory of relativity, although he was responsible for many other important results as well.

4.13.A   Brief Biographical Sketch

Einstein was born in Germany on March 14, 1879. He was thought to have little scholastic ability, and in addition, was believed to be dyslectic. He had no interest in physical activity or sports. His father and uncle had set up an electrical and electrical engineering plant in Munich, a Catholic city. Albert was the only Jewish boy in the local school. But Albert's family, not being religious, was not a part of the local Jewish community. As a consequence, Albert was not to come to grips with his ethnic identity until many years later. By age 15 the family business failed and the family moved south to Milan, Italy, but without Albert; he remained behind to try to complete his schooling. But six months later, Albert, who resented the regimented format of Teutonic schooling and who had poor grades in geography, history, and languages, quit his secondary school studies in Munich without having received any diploma and joined the family in the more relaxed environment of Milan.

In 1896, while thinking about continuing on with his education to college, Einstein's distaste for Germany compelled him to renounce his German citizenship, despite the fact that he then became stateless. It was difficult to continue on to college without a formal high school diploma from Munich. So Albert decided to apply to the renowned Federal Polytechnic School in Zurich, Switzerland, which did not require a diploma but where he would have to pass an entrance exam. He took the exam but failed. However, by doing preparatory studies in Switzerland for a year he finally passed the entrance examination. At the Polytechnic he focused on studies in mathematics and physics rather than on electrical engineering, as his family would have preferred, and graduated in 1900 at age 21. In 1901, he became a Swiss citizen.

While in Zurich, Albert enjoyed the company of a number of women. One such during his college years was Mileva Maric, a mathematics student of Hungarian origin who was four years his senior. They became sweethearts and married in 1903. Einstein's joy in women and his need to have them in his life were to be a characteristic throughout his life.

After graduation in 1900, Albert was unable to find employment. He finally found a two-month job as a mathematics teacher. Through friends, however, he was referred to the Swiss patent office in Bern, Switzerland, in June 1902 for a job as an examiner. He remained at the patent office until 1909. During the period 1902–1905, since Einstein was married, had the security of a guaranteed income because of his government job at the patent office, and had finalized his citizenship, his life was stable. He then had time to work, in his spare time, at his principal interest and passion, physics. Not yet in the academic world, and not yet required to do research for his job, Einstein was carrying out his research from purely personal curiosity.

By 1904, Einstein had already published five papers, mainly on the theory of heat. But the excitement was yet to come. The year 1905 was to be a banner year for 26-year-old Albert Einstein; he published four separate scientific papers, each of which was to open up a distinct subfield of physics. One of the four dealt with the quantification of the theory of Brownian motion of small particles suspended in fluids. In that paper, Einstein showed how to calculate the masses and sizes of atoms, and it was for that paper that he received his doctorate from the University of Zurich in 1905. For these four 1905 papers, and another published in 1906, Einstein would eventually be acclaimed throughout the world for the genius which revolutionized the field of physics and changed the world's view of space and time forever.

In the winter of 1908, at the age of 29, while still employed at the patent office, he began his academic career by lecturing to audiences of two to four students at the University of Bern. A year later, he left the patent office for a full-time associate professorship at the University of Zurich. By 1911 he had left Zurich to teach at the German University in Prague, Bohemia (now the Czech Republic). But life in Prague, aside from his work, was difficult. He was a man with a Germanic background in a Slavic country and did not speak Czech. He found himself confined to the German community of Prague, while his Slavic wife became involved in the Serbian struggle for independence. Einstein himself had no interest in such a cause and the couple began to drift apart. In August 1912, Albert returned to the University of Zurich as a full professor; in this move back to Switzerland he was accompanied, for the last time, by his wife and children.

In the spring of 1914, Einstein accepted a position with the Prussian Academy of Sciences in Berlin. He decided to return to Germany despite his many negative feelings about the country and despite increasing talk of the possible outbreak of war, because of the opportunities there would be in Berlin to interact with some of the best scientists in Europe and because of the high salary promised him. There he could carry out his research unencumbered by administrative and teaching duties, and occasionally, he could give lectures at the very prestigious University of Berlin.

By this time, the Einsteins had two sons, Hans Albert and Eduard. When Albert moved to Berlin, his wife and sons remained behind to vacation in Switzerland. But in August 1914, the outbreak of World War I prevented his family members from joining Albert in Berlin. This forced separation finally precipitated a more formal marital separation; Albert and Mileva remained separated until the war ended in 1918, at which point they officially divorced. Albert then married his widowed cousin Elsa in the summer of 1919.

Einstein was an outspoken critic of German militarism, and openly adopted a pacifist position, one he would hold through the remainder of his life. During World War I he continued his research into the general theory of relativity and enjoyed a decade of research accomplishment that was comparable in importance to his banner year of 1905. Einstein had suggested that his new general theory of relativity might be tested empirically by observing the precession (wobbling) of the orbit of the planet Mercury during an eclipse. The eclipse required to test Einstein's general relativity theory was to occur in 1919. Scientists traveled to New Guinea to make measurements, and Einstein's predictions about the orbit of Mercury were verified precisely (within experimental error). This confirmation finally brought international fame to Einstein. Up until then, his work was known mostly only to other scientists. But from that time on, the name of Einstein became a household word in general society, synonymous with genius. A very smart person was called "an Einstein."

Ironically, as Einstein's reputation grew outside Germany, he came under increasing attack within Germany because he was Jewish. Anti-Semitism was spreading throughout Germany as the society searched for reasons why it had lost World War I. Postwar Germany, looking for a scapegoat, blamed the Jews. Einstein was attacked by both German scientists, who claimed that his relativity work was all wrong, and by German society in general. Up until then Einstein had hardly known that he was Jewish; now he had to defend himself against blind bigotry. In 1919, Einstein attended a conference in Berlin on Zionism, a meeting called to discuss the problems and future of the Jewish people in the Diaspora. The conference had a powerful effect upon Einstein; he became an ardent Zionist and would be involved in Zionist affairs for the rest of his life. In 1921, at Chaim Weizmann's request, Einstein toured the United States to raise money for the Palestine Foundation Fund.

In 1921, Einstein was awarded the Nobel Prize for Physics for his research on the photoelectric effect and for general contributions to theoretical physics. His photoelectric effect work was published in one of his 1905 papers. His work on relativity, controversial at the time, was not specifically mentioned. Einstein remained in Berlin at the Prussian Academy of Sciences for 19 years, until 1933. During this period he worked on trying to relate electromagnetism to gravitation and thus find a unified field theory relating all the forces of nature. (Faraday had attacked the same problem with the same negative result; see Section 4.8.)

In 1933, Adolph Hitler became chancellor of Germany. Realizing immediately the implications for his survival, Einstein left Germany as soon as possible. He was taken by private yacht from Belgium (where he had been lecturing) to England, and then he traveled to the United States, where he had accepted a job at the newly established Institute for Advanced Study in Princeton, New Jersey. Einstein's wife Elsa died in 1936. He was left a very lonely man, alone in yet another new country.

Einstein was to become involved in World War II in his newly adopted country. He watched the rise of Adolph Hitler in Germany from afar and followed the systematic persecution of the Jewish people in the nation where he had spent most of his life. He considered the members of the Jewish community in Germany to be fools for trying to be loyal to Germany and not appreciating what fate the Nazi regime had in mind for them. In a famous letter dated August 2, 1939, one month before the beginning of World War II, Einstein advised President Roosevelt that ". . . the element uranium may be turned into a new and important source of energy in the immediate future. Certain aspects of the situation which has arisen seem to call for watchfulness, and if necessary, quick action on the part of the administration. . . ." Einstein suggested that an atomic bomb might be carried into a port by a boat and exploded, destroying the port and much of the surrounding territory. Roosevelt consequently appointed administrators and scientists who carried out the development of the atomic bomb that ended the World War II in 1945. Einstein was to have misgivings about the atomic monster that he had suggested be released. He felt partly to blame for the horror inflicted upon the Japanese people by the dropping of atomic bombs on Hiroshima and Nagasaki.

In 1948, the United Nations recreated the Jewish State of Israel after a gap of 3000 years. Einstein was offered the presidency of the new nation in 1952. He turned it down, claiming incompetence for the task. Einstein remained at the Institute for Advanced Study in Princeton during the final 22 years of his life, and died in 1955 at the age of 76. He was never successful in finding a unified field theory that was acceptable to the scientific world. On April 6, 1922, in an address to the French Philosophical Society at the Sorbonne, in Paris, Einstein said: "If my theory of relativity is proven successful, Germany will claim me as a German, and France will declare that I am a citizen of the world. Should my theory prove untrue, France will say that I am a German, and Germany will declare that I am a Jew."

4.13.B Einstein's Scientific Contributions

In 1905, Einstein published four fundamental research papers:

1. A theoretical explanation for Brownian motion based on thermodynamics and statistical mechanics
2. A new representation of light in terms of particles called *quanta*, or *photons*, that gave rise to the photoelectric effect (light could then be viewed as particles instead of waves)
3. The special theory of relativity
4. An equivalence of mass and energy

Together with the general theory of relativity, these papers were his major scientific contributions; each was revolutionary and changed the way we view the universe.

Brownian Motion (1905)

In 1828, Robert Brown, a Scottish naturalist, had discovered that when pollen dust (particles of the order of 0.001 millimeter) is suspended in water, the specks of dust exhibit a continuous, zigzag, and seemingly random motion. Brown found that the same random motion, now known as *Brownian motion*, resulted when small particles of either organic or inorganic matter were suspended in any liquid. But Brown provided no explanation for this phenomenon. In 1905, Einstein provided an explanation and quantified the effect. What fascinated Einstein was that the motion seemed contrary to all experience, in that it never diminished although no external forces seemed to be acting on the particles. His explanation of the motion was that it was due to the kinetic energy of the invisible molecules with which the particles were constantly colliding. He then went on to predict the mass and number of molecules involved and thereby quantified the nature of the Brownian motion. His predictive law was probabilistic and given by the "normal" or bell-shaped curve. It stated that the probability distribution for the distance that a particle suspended in a fluid would move in any given direction in a given time interval is given by the normal probability law in which the variance is proportional to the length of the time interval.

Einstein asserted statistically that the mean kinetic energy of agitation of the particles was expected to be exactly the same as the energy of agitation in a fluid molecule (i.e., a liquid or a gas). By showing their impact on suspended particles, Einstein had demonstrated the existence of the invisible molecules; the existence of molecules is a fact we take for granted today. He had provided a basis for prediction, based on subjective hypothesizing and theory alone; his predictive law was not developed on the basis of experimentation. That approach to science was the one that Einstein was to use in all his research. It was an approach based on mathematical formulation of laws of nature, formulations whose implications could be studied.

Photoelectric Effect (1905)

In 1900, Max Planck had asserted that energy is radiated not in a continuous flow but in discrete bursts or quanta (photons). The amount of energy radiated is proportional to the frequency of the associated wave of electromagnetic radiation. For example, violet light, which has twice the frequency of red light, would emit twice the number of photons per unit time. So there seemed to be a conflict between beliefs about the very nature of light. Did light consist of waves or particles? Or was there a duality in the nature of light?

Working with Planck's explanation of how light is constituted, Einstein postulated a physical law to explain what happens when light is shined on matter. He explained that light can be wavelike but behave like flows of discrete particles. He reasoned that when electromagnetic radiation impinges on matter, each photon in the radiation penetrates the matter and collides with an atom. Each atom has electrons, and when the incoming photon hits the atom, the atom ejects an electron, which is given great speed by the interaction. When many photons interact with the matter, many electrons are released and a current of electricity flows. Einstein showed that the maximum kinetic energy of the ejected electron is proportional to the frequency of the radiation being used to stimulate the action, diminished by the work done by the electron in escaping from the surface of the matter. The number of ejected electrons (although not their speed) varies with the intensity of the incident light. The kinetic energy of the ejected electrons is proportional to the voltage produced by shining the light on the matter (usually, a metal). So the size of the electric current to be expected can be calculated from Einstein's mathematical formulation. For his explanation of this phenomenon, which came to be called the *photoelectric effect*, Einstein won a Nobel prize in 1921. The applications of this work are found, for example, in the development of television, the light meter, the laser, the videotelephone, the computer screen, talking motion pictures, and videos.

Special Theory of Relativity (1905)

In 1887, Albert Michelson and Edward Morley carried out a famous experiment that was designed to study the effect on the speed of light of Earth's passage through a postulated *ether* that surrounds it. It had been thought since the time of the Greeks that there had to be a medium (the ether) through which electromagnetic radiation (including light) had to travel, just as sound requires molecules of air to vibrate before it can be heard. It was reasoned that if a man walks forward along the deck of a ship at the rate of 1.5 meters per second, and the ship moves through the water in the same direction at the rate of 7.5 meters per second, the man's speed with respect to the water is 9 meters per second. If he walks aft, his speed relative to the water is 6 meters per second. If he could walk at the same speed at which the ship is moving through the water, but in the opposite direction, and if the water were not in motion, an observer on land would perceive that the man was not moving at all.

The same phenomenon of addition and subtraction of speeds is observable with the sound of a bell ringing at a railroad crossing when heard by someone on an approaching or departing train. If the train is traveling at 25 meters per second and if the sound is traveling at 400 meters per second, the speed of the sound relative to the train is 425 meters per second when the train is approaching the crossing, but 375 meters per second when the train is leaving the crossing. When this same reasoning was applied to the speed of light, serious problems arose after the Michelson–Morley experiment was carried out. We quote from Barnett (1948, p. 47):

> Specifically, a light ray projected in the direction of the earth's movement should be slightly retarded by the ether flow, just as a swimmer is retarded by a current when going upstream. The difference would be slight, for the velocity of light . . . is 186,284 miles a second, while the velocity of the earth in its orbit around the sun is only 20 miles a second. Hence, a light ray sent *against* the ether stream should travel at the rate of 186,264 miles a second, while one sent *with* the ether stream should be clocked at 186,304 miles a second. Michelson and Morley constructed an instrument of such great delicacy that it could detect a variation of even a fraction of a mile per second in the enormous velocity of light (the instrument was called an interferometer). . . . And the result was simply this: there was no difference whatsoever in the velocity of the light beams regardless of their direction.

That is, the experiment did not find any ether. It was concluded that there is no such medium. The implications for physics were enormous. Light propagation from, or to, Earth would be unaffected by an ether, since such a medium did not exist. According to Barnett (1948, p. 49): "The one indisputable fact established by the Michelson–Morley experiment was that the velocity of light is unaffected by the motion of the earth. Einstein seized on this as a revelation of universal law. If the velocity of light is constant regardless of the earth's motion, he reasoned, it must be constant regardless of the motion of any sun, moon, star, meteor, or other system moving anywhere in the universe. From this he drew a broader generalization, and asserted that the laws of nature are the same for all uniformly moving systems. This simple statement is the essence of Einstein's Special Theory of Relativity."

If there is acceleration, the *general theory of relativity* would be invoked, but for constant (uniform) speed, *special relativity theory* would be sufficient for an explanation. The implications of the special theory were very difficult for the nonspecialist to grasp. For example, since speed is the ratio of distance divided by time, if speed (of light) is to remain constant as light travels over different distances, this ratio must remain constant, and the only way that can happen is if the concepts of distance and of time are both modified. To accomplish this constancy of the speed of light, Einstein demonstrated mathematically precisely how the classical notions of distance and time had to be modified. He showed, for example, that an object in uniform motion relative to an observer would have to have its length modified relative to its length when at rest. In fact, it would lengthen by a factor that depends on the square of the ratio of its speed to the speed of light. He showed using the Lorentz transformation equations (which had been developed for a different purpose) that similar relationships would have to hold for time itself and for the mass of the object. The Lorentz transformation equations (Barnett, 1948, p. 58) "preserve the velocity of light as a universal constant but modify all measurements of time and distance according to the velocity of each system of reference."

Events in any two frames of reference could be connected in terms of these specialized Lorentz transformation equations, instead of in terms of the Newtonian transformation equations that had served well for so long. The result for the object in motion, say a rod, was that its length would shrink or contract; elapsed time would be reduced (clocks would run more slowly); and mass would grow (the mass of the rod in uniform motion would increase from its rest mass value). The effects would be noticeable only for objects traveling at speeds that were significant fractions of the speed of light, and for those, the effects could be substantial. Otherwise, results obtainable from Newtonian mechanics would be satisfactory as an approximation.

Equivalence of Mass and Energy (1905)

Einstein's fourth 1905 paper was, in a sense, an extension of the special theory of relativity. The paper provided a formal way of relating the energy of a radiating object to its mass. As the object radiates by emitting light rays, it loses mass, but how much? This question resulted in arguably the most famous equation of all time: $E = mc^2$—the energy of the radiating object is proportional to its mass, and the proportionality constant is the squared speed of light (c^2). This equation was used, for example, to predict the amount of energy that would be released in the atomic bomb explosions of 1945.

General Theory of Relativity

The general theory of relativity, developed during the decade following the 1905 development of the special theory of relativity, produced many other mathematical predictions. For example, light passing from one part of a gravitational field to another would be shifted in wavelength. This effect has been referred to as the *Einstein red shift*. Just as he had concluded for the special theory of relativity, Einstein asserted for the general theory that the laws of nature are the same for all systems, regardless of their state of motion. The playing out of this premise involved Einstein in modifying the laws of gravitation. He proposed the *principle of equivalence of gravitation and inertia.* That is, there is no way to distinguish the motion produced by inertial forces (acceleration, centrifugal force, recoil, etc.) from motion produced by gravitation. For Newton, gravitation was a *force*; for Einstein, it was a *field*, as Faraday and Maxwell viewed the field of electricity around a magnet. Einstein concluded that like any material object, light travels in a curved path when traveling through the gravitational field of a massive body.

Using the general theory of relativity, Einstein predicted that light from stars surrounding the Sun should be bent inward, toward the Sun, in passing through the Sun's gravitational field. So the images of those stars should be shifted outward from their usual positions in the sky. For the closest stars to the Sun, he predicted that the deviation should be about 1.75 seconds of arc. When the eclipse of May 29, 1919 was observed in equatorial latitudes, Einstein's predictions turned out to be confirmed exactly, within limits of measurement accuracy. This result astonished both the scientific world and society in general and brought Einstein universal acclaim.

4.13.C   Major Works

Scientific Papers

Most of Einstein's scientific work is contained in the papers listed in Ernst Weil, Ed. *Albert Einstein: A Bibliography of His Scientific Papers, 1901–1930.* London: 1937.

Albert, Einstein (1905). "Uber Einen die Erzeugung und Verwandlung des Lichtes Betreffenden Heuristischen Gesichtspunkt," *Annalen der Physik*, Ser. 4, **17**:132–148.

———(1905). "Uber die von der Molekularkinetischen Theorie der Warme Geforderte Bewegung von in Ruhenden Flussigkeiten Suspendierten Teilchen," *Annalen der Physik*, Ser. 4, **17**:549–560.

———(1905). "Zur Elektrodynamik Bewegter Korper," *Annalen der Physik*, Ser. 4, **17**:891–921.

———(1905). "Ist die Tragheit eines Korpers von Seinem Energieinhalt Abhangig?" *Annalen der Physik*, Ser. 4, **18**:639–641.

———(1906). "Zur Theorie der Brownschen Bewegung," *Annalen der Physik*; translated separately as *Investigations on the Theory of the Brownian Movement.* London: Methuen and Co. 1926.

———(1907). "Plancksche Theorie der Strahlung und die Theorie der Spezifischen Warme," *Annalen der Physik*, Ser. 4, **22**:180–190, and p. 800 (Berichtigung).

———(1911). "Uber den Einfluss der Schwerkraft auf die Ausbreitung des Lichtes," *Annalen der Physik*, Ser. 4, **25**:898–908.

————(1913). "Entwurf einer Verallegemeinerten Relativitatstheorie und einer Theorie der Gravitation," *Zeitschrift fur Mathematik und Physik*, **62**:225–261.

————(1916). "Die Grundlage der Allgemeinen Relativitatstheorie" (The Foundation of the General Theory of Relativity), *Annalen der Physik*, Ser. 4, **49**:769–822.

————(1917). "Quantentheorie der Strahlung," *Physikalische Zeitschrift*, **18**:121–128.

————(1924). "Quantentheorie des Einatornigen Idealen Gases," *Sitzungsberichte der Preussischen Akademie der Wissenschaften* (1924 and 1925), 261–267.

————(1950). *The Meaning of Relativity, Including the Generalized Theory of Gravitation*, 3rd ed. Princeton, NJ: Princeton University Press.

Lorentz, H. A., Albert Einstein, H. Weyl, and A. Minkowski (1923). *The Principle of Relativity: A Collection of Original Memoirs on the Special and General Theory of Relativity*. London: Methuen and Co.

Albert Einstein (1987) Stachel, John, et al. (Eds.). Princeton, NJ: Princeton University Press. *The Collected Papers of Albert Einstein* (Contains all his papers, notes, and letter, with companion translation volumes.)

Other Works

Einstein, Albert (1930). *About Zionism: Speeches and Letters*, English translation by Sir Leon Simon. New York: The Macmillan Co.

————(1950). *Out of My Later Years*, paperback ed. New York: Wisdom Library of the Philosophical Library.

Einstein, Albert, and Sigmund Freud (1932). *Builders of the Universe*. 1933, "Warum Krieg?" (Why War?), English translation (1978) by Fritz and Anna Moellenhoff Chicago: Chicago Institute for Psychoanalysis.

Einstein, Albert (1949). *The World as I See It*. English translation by Alan Harris. New York: Philosophical Library.

Einstein, Albert, and Leopold Infeld (1938). *The Evolution of Physics, from Early Concepts to Relativity and Quanta*. New York: Simon and Schuster.

4.13.D   Subjectivity in the Work of Einstein

Albert Einstein was a theoretician, not an experimentalist. His research did not follow the classical scientific tradition, established from the time of Galileo, of hypothesizing and then designing experiments to test the hypothesis. Instead, Einstein's methodological approach was to make some assumptions, develop mathematical representations of the assumptions, and then to study, mathematically and physically, the implications of the structural models built from the assumptions. Einstein would then carry out what he called *gedanken experimente*, thought experiments that would test his ideas theoretically, experiments possible in theory but ruled out by experimental difficulties. For example, he might create an "experiment" in which an observer is in an elevator traveling at near the speed of light, and then he would speculate about what the observer might see happen to a ball thrown up in the air in the elevator. In one of Einstein's later writings he tells the story that at age 16, he wondered what would happen if one could follow a beam of light at the

speed of light. He concluded that the observer would be in a "spatially oscil-
latory electromagnetic field at rest." In the *twin paradox*, Einstein carried out
a *gedanken experiment* in which one identical twin stayed home but the other
traveled through the universe at great speed. Einstein concluded that they
must age differently. The one traveling through the universe at great speed
ages slowly, returning home only to find that his twin has been dead for
perhaps hundreds of years.

But where did Einstein's assumptions and hypotheses come from?
Ronald W. Clark has commented (1984, p. 54): "Thus it was physics to which
he turned, working 'most of the time in the physical laboratory, fascinated
by the direct contact with experience' [Schilpp, p. 15]. This contact with
experience was in strange contrast with the period when he would answer
a question about his laboratory by pointing to his head, and a question
about his tools by pointing to his fountain pen. Yet despite this he never
ceased to emphasize that the bulk of his work sprang directly and naturally
from observed facts; the coordinating theory explaining them might arise
from an inspired gleam of intuition, but the need for it arose only after
observation."

Einstein pointed out that his formulation of the special theory of relativity
"was not speculative in origin; it owes its invention entirely to the desire to
make physical theory fit observed facts as well as possible" (Lecture, Kings
College, London, 1921, quoted in *The Nation*, London, June 18, 1921). When
Einstein was studying and trying to quantify Brownian motion, he provided a
virtual proof for the existence of molecules, particles invisible to the human
eye whose existence was postulated by theory rather than produced by experimental evidence. Einstein thus believed that theories into which facts were
later seen to fit were more likely to stand the test of time than theories constructed entirely from experimental evidence.

According to Clark (1984, p. 118), "Einstein was always ready to agree that
inventiveness, imagination, the intuitive approach—the very stuff of which
artists rather than scientists are usually thought to be made—played a serious
part in his work." Einstein, himself said (recalled by a friend on the one hundredth anniversary of Einstein's birth, celebrated February 18, 1979, quoted
in Ryan, 1987): "When I examine myself and my methods of thought I come
to the conclusion that the gift of fantasy has meant more to me than my talent
for absorbing positive knowledge."

Einstein had very strong subjective convictions about the assumptions he
made that were to form the basis of his theoretical constructs. For example, in
a letter of March 17, 1942 to one of Albert Michelson's biographers, he said:
"It is no doubt that Michelson's experiment was of considerable influence
upon my work insofar as it strengthened my conviction concerning the validity of the principle of The Special Theory of Relativity. On the other side, I
was pretty much convinced of the validity of the principle before I did know
this experiment and its result. In any case, Michelson's experiment removed
practically any doubt about the validity of the principle in optics and showed

that a profound change of the basic concepts of physics was inevitable." Einstein claimed that he had learned about the Michelson–Morley experiment only *after* 1905; otherwise, he claimed, he would have included reference to it in his 1905 paper. Einstein told R. S. Shankland of Case Institute, on February 4, 1950: "I am not sure when I first heard of the Michelson–Morley experiment . . . I just took it for granted that it was true."

David Ben-Gurion once asked Einstein whether the theory of relativity was the result of his thought processes only. Einstein confirmed that it was, but added (*Jewish Observer*, Feb. 22, 1955): "I naturally had before me the experimental works of those preceding me. These served as material for my thoughts and studies." R. W. Clark (1984, p. 222) pointed out: "Only Einstein the philosopher could have convinced Einstein the scientist that if the evidence did not agree with the theory then the evidence must be faulty."

In his 1911 paper on gravitation, Einstein tried to answer the question of whether the propagation of light is influenced by gravitation. Was light bent by gravity as it passed near the sun? From 1911 onward he pointed out with increasing persistence that here was one way of proving or disproving experimentally a theory which had been built up logically, but which had as its foundation, little more than an intuitive hunch (Clark, 1984, p. 183). In a note to Carl Seelig (Seelig, 1956, p. 188) Einstein wrote: "I concentrated on speculative theories, whereas Rutherford managed to reach profound conclusions on the basis of almost primitive reflection combined with relatively simple experimental methods."

Max Born, on reading Einstein's 1916 paper "The Foundation of the General Theory of Relativity," wrote (Born, 1916, p. 253) that it was "the greatest feat of human thinking about nature, the most amazing combination of philosophical penetration, physical intuition, and mathematical skill. But its connections with experience were slender. It appealed to me like a great work of art, to be enjoyed and admired from a distance."

Einstein generally carried out his research with a strong inner belief that bordered on certainty that he was right. He somehow had very strong preconceived notions about what concepts in physics had been formulated improperly and how they should be reformulated. On any given topic, he was able first to search the extant knowledge base and then carefully scrutinize it for any inconsistencies and unreasonable implications. When he encountered such an unreasonable inconsistency, he took on the task of trying to reformulate the physical relationships in ways that would eliminate the problems with previous theory. He held strong personal beliefs about what should be expected in the physical world and then set about structuring mathematical laws and descriptive systems that would lead to his anticipated conclusions. He was extremely subjective in his approach to scientific research, perhaps more subjective than any other scientist had ever been.

REFERENCES

Barnett, Lincoln (1948). *The Universe and Dr. Einstein.* New York: New American Library of World Literature, Mentor Books.

Born, Max (1916). *Physica Acta.* Hebrew University, Israel: The Einstein Archives, p. 253.

Calaprice, Alice (1996). *The Quotable Einstein.* Princeton, NJ: Princeton University Press.

Clark, Ronald W. (1984). *Einstein: The Life and Times.* New York: Avon Books.

Einstein, Albert (1993). *Einstein on Humanism.* Secaucus, NJ: Carol Publishing Group, Citadel Press.

Frank, Philipp (1947, 1953). *Einstein: His Life and Times* (English translation). New York: Alfred A. Knopf.

Hoffman, Banesh (with the collaboration of Helen Dukas) (1972). *Albert Einstein: Creator and Rebel.* New York: Viking.

Holton, Gerald (1996). *Einstein, History and Other Passions.* Reading, MA: Addison-Wesley.

Michelmore, Peter (1962). *Einstein: Profile of the Man.* New York: Dodd.

Regis, Edward (1987). *Who Got Einstein's Office.* Reading, MA: Addison-Wesley.

Ryan, Dennis P. (Ed.) (1987). *Einstein and the Humanities.* New York: Greenwood Press.

Schilpp, Paul A. (Ed.) (1949). *Albert Einstein: Philosopher–Scientist,* 2nd ed. Evanston, IL: Library of Living Philosophers.

Seelig, Carl (1956). *Albert Einstein: A Documentary Biography.* London: Staples Press.

Vallentin, Antonina (1954). *Einstein: A Biography* (English translation). Garden City, NY: Doubleday; Paris: Plon et les Librairies Associés.

4.14 SOME CONJECTURES ABOUT THE SCIENTISTS

We have now completed our task of recounting a story for the life and scientific methodology of each of the 12 scientists listed in Meadows (1987) *The Great Scientists.* We believe that our own strong prior belief that we would find generous use of subjectivity in the scientific work of each of the 12 is confirmed. It might be of interest to conjecture a bit about what the famous scientists might have done had Bayesian analysis been available to them. Suppose that advanced mathematics, subjective probability, and Bayesian statistical methods of analyzing scientific data had already been developed and were readily available at the time some of the scientists we have studied lived and worked. How might that research methodology have been employed to strengthen the positions that those scientists held, thus perhaps settling more definitively some of the questions still raised, even today, perhaps hundreds of years later? In many cases we are still questioning their research methodologies, their judgments, their conclusions, and their decision making that was based on the conclusions they reached. In the following summaries of the subjectivity of each of these scientists, we shall speculate in some cases as to how they might have used Bayesian analysis. In particular, we shall speculate on how they might have used such analysis to compare competing models of the phenomena under investigation. (We show in Chapter 5 how to use Bayesian analysis to compare scientific models.)

Aristotle was a dedicated and careful observer of the natural world. Nevertheless, his belief in final causes and other overarching theories sometime led him to overgeneralize or otherwise distort his observations. For example, he wrote that human beings are superior to other animals in purity of blood and softness of flesh. Because he believed that males are better equipped than females with offensive and defensive weapons, he concluded that worker bees are male (when, in fact, they are female). He overgeneralized his observation that extremely light objects fall more slowly than do heavy ones to conclude that the speed of falling is generally related to the weight of an object. (We now understand that the difference in speed between extremely light and heavy objects occurs only because the effect of air resistance is more obvious for very light objects.) He described detailed observations that purport to show, mistakenly, that the heart is the first part of the embryo to develop. These mistaken observations lent support to his doctrine that the heart is the principle of life, the seat of the soul, of locomotion, and of what we now call higher mental functions. Since Aristotle was more inclined to speculate about scientific facts than to collect data to support his conclusions, we shall not conjecture in his case about how he might have utilized probability.

Galileo seems to have been convinced a priori of the law of motion that the distance fallen by a dropped object is proportional to the square of the time elapsed from dropping well before he conducted experiments to demonstrate the law. Indeed, he seems to have derived the law from a set of false premises. Further, there is some doubt as to whether the instrumentation of his day would have permitted Galileo to carry out the experiments he claims to have done, or at least to have achieved results of sufficient accuracy to prove his thesis, had he not already been convinced of its truth. Despite his position that seemed to challenge the Church's insistence on a geocentric universe, Galileo's own belief in cosmic order prevented him from accepting Kepler's substitution of elliptical planetary orbits for the more cosmically perfect circular ones. This rejection barred Galileo from full formulation of the inertial law, eventually formulated by Newton.

Galileo was forced to entertain two opposing theories of how Earth travels through the universe: Does Earth travel around the Sun, or does the Sun travel around Earth? Copernicus and Kepler had already concluded that the heliocentric theory was to be preferred to Ptolomy's geocentric theory. Galileo's professional career was dominated by his life-threatening confrontation with the Catholic Church over which of these two theories should be accepted as truth. Putting political and religious considerations aside, Galileo might have calculated the (posterior) odds favoring each theory from a Bayesian viewpoint by comparing the (subjective) probabilities of each of the two theories to see how they compared (the methodology required for accomplishing this is explicated in Chapter 5). The powerful political position of the Church at that time, and its strong and long-held fixed belief in geocentrism, suggest that even if such a calculation had yielded results that showed the probability of the heliocentric theory overwhelming the probability of the geocentric theory, it still would have been difficult to convince the Church authorities.

William Harvey used what he called a "meditation" on the amount of blood issuing from the heart over a period of time to deduce that such a large quantity could not be manufactured or stored in the body; hence he was led to the notion of circulation. Having convinced himself of the reality of circulation and thus the necessity for blood to flow from arteries to veins, he was willing to believe (correctly) in the existence of a connecting network of capillaries without ever having physically observed such a network. In his work on the reproductive mechanisms of animals, Harvey was unable to identify a human ovum but was sufficiently confident of the generality of his findings in lower animals that he subjectively (and correctly) generalized them to the entire animal kingdom.

Harvey confronted Galen's earlier theory about the heart, aorta, and all the arteries—that the blood pulsates through the body in an ebb-and-flow pattern, and certainly doesn't travel in a closed, one-directional path—by examining the implications of such a theory. He actually made calculations about what such a pulsation theory would imply about the buildup of blood in various places in the body, and he collected data that would bear on both the pulsation and circulation hypotheses. Had Bayesian methods of analysis been available, he might have formed the posterior odds ratio for the probability of the pulsation model for blood compared with the probability of the circulation model he was considering. A ratio much less than 1 would have argued very strongly for the circulation model, and might have been very convincing to Harvey's colleagues. Harvey might also have applied such a probabilistic approach to comparing his belief in the old Aristotelian theory of epigenesis (the formation of a fetus by the addition of one part after another) with the more commonly held thinking of his time, the theory of preformation (all parts of the animal fetus were present, if invisible, from the start). His data, obtained from numerous dissections, strongly supported his belief.

Subjectivity seems to have entered the work of Sir Isaac Newton at a later point in the scientific process. Between the first and second editions of the *Principia*, Newton has been found to have made several changes to make the data he reported seem to fit more closely the theories he was advocating. He reported a calculated value of the speed of sound that agreed too exactly with a value reported by another investigator, despite the fact that the earlier value was the average of many distinct measurements. Newton claimed that water vapor constitutes 10% of air (in fact, it does not, but varies with temperature, pressure, and geographic location). He further claimed, without empirical support, that sound is not propagated through water vapor. To compensate,

Newton arbitrarily increased the calculated speed of sound by 10%. Numbers purported to be data on the precession of the equinoxes were altered to better fit some corrected mathematics. Because Newton was sure of the truth of his theories, he saw his task as that of convincing his peers, and hence sometimes altered his experimental results to make them seem to support the theory more strongly. But excluding his work in alchemy and other metaphysical researches, he got his physical laws right (although they were somewhat modified later by Einstein).

A major issue of model comparison concerned Newton during most of his career, the issue of the physical nature of light. Should light be thought of as corpuscular (made up of tiny particles) or as a wave motion? Newton was mostly persuaded during his early years that light should be thought of as a wave motion (and he studied the interference patterns now called Newton's rings, which invoked wave-motion thinking). Later in his career he switched positions and then thought about light as made up of particles whose behavior could be predicted by his theory of the laws of motion of material bodies. In this matter he was to be pitted against Christian Huygens, another famous scientist of the time, who strongly believed in the wave theory of light. The issue remains an arguable matter today. In what is called the *dual nature of light* (a theory partially attributable to Max Planck), light sometimes seems to act one way and sometimes the other. Would a Bayesian model comparison have helped? We believe it might have, and might still, but such a probability comparison does not yet seem to have been carried out.

For Antoine Lavoisier, there is a sense in which his entire corpus of work, consisting as it did primarily of a theoretical synthesis of the experimental work of others, constitutes an exercise in subjectivity. He spent a good part of his career refuting the phlogiston theory (the mistaken theory that all flammable substances contain something called phlogiston that is given off when they burn). Indeed, he was the originator of the antiphlogiston theory of combustion. Yet he was not above invoking explanations drawn from phlogiston when these seemed convenient, and he advocated the imponderable "caloric" to help explain the varying states of matter. Not only may Lavoisier have plagiarized the experiments of others, claiming to have repeated those experiments but perhaps not actually having done so, but experiments that he reported having done were sometimes carried out after the publication in which they were cited. Further, the accuracy of experimental results was sometimes exaggerated, and the number of replications achieved similarly inflated. Like Newton, Lavoisier was subjectively convinced of the accuracy of the scientific system he was proposing, the systematization of chemistry, and sometimes gave in to the temptation to embellish the empirical support for that system. But the system and most of its details, especially the understanding of combustion and other forms of oxidation, were sufficiently correct to earn Lavoisier the sobriquet "father of modern chemistry." The argument between the antiphlogistians and the advocates of the phlogiston theory of combustion might have been resolved more readily had the posterior probability calculations been presented to researchers caught up in Lavoisier's chemical revolution. Again, the issue was selection of the correct model to use to come closer to truth.

Early in his scientific career, Alexander von Humboldt set out to demonstrate the harmoniousness of nature; accordingly, his voluminous writings repeatedly make the point that nature is one great whole, in which plants, animals, and geological and meteorological phenomena, as well as human beings and their culture, fit coherently. Von Humboldt's subjectivity influenced him, early in his career, to interpret geological observations as evidence supporting the Neptunist theory of the origin of rocks in sedimentation. Later, more extensive observations and a change in his theoretical orientation led to a reversal of the interpretation of the same observations as evidence in favor of the volcanic origin of the rocks (the Plutonist theory). Bayesian statistical methods of analysis, had they been available at the time, could readily have compared the Neptunist and Plutonist theories.

Michael Faraday, like von Humboldt, sought unity in nature. Under the influence of his religion of Sandemanianism, Faraday believed in the simplicity and integration of nature. This led him to attempt to establish the unity of the forces of magnetism, electricity, and gravitation. He achieved the conversion of electrical to mechanical energy and established the equivalence of all types of electricity. His conviction led to his persistence through many years of attempts before he was successful in showing the conversion of magnetism to electricity. He continued to believe in the relationship between electricity and gravity, although he was unable to establish that relationship. Faraday even left a letter to be opened 60 years after his death that carried the idea of unity of forces further with speculation on the existence of electromagnetic waves analogous to those of the ocean.

Charles Darwin believed theory must guide observations despite his professed belief in Baconian induction. His synthesis of his observations led him to a theory of evolution powered by a mechanism of natural selection yielding gradual changes in organisms. This subjective belief led him to explain gaps in the fossil record, where intermediate forms should have been represented, as being merly temporary gaps in data gathering, to be filled as paleontologists explored further. It also led him to disbelieve then-current calculations of the age of Earth, since the accumulation of gradual changes would require more time than those calculations provided. Scientific opposition to Darwin's proposed mechanism of natural selection was coupled with the mistaken belief (which Darwin shared) that the material of heredity was "infinitely divisible." If that were the case, the substance in the human body that passed heredity characteristics such as eye color on to the next generation could pass on any degree of "blueness," for example, as opposed to just "blue" or "not blue." An implication of this mistaken belief was that a change (mutation) in an organism would be diluted through interbreeding. That implication caused Darwin to change his mind about natural selection. His later writings gave but scant importance to natural selection and stressed instead a purely subjective notion of pangenesis, which implied the inheritance of acquired characteristics. Neither Darwin nor any of his contemporaries or successors could find any empirical evidence for the inheritance of acquired characteristics, called the *Lamarckian view*. The lack of empirical support for Lamarckian mechanisms was coupled with the rediscovery of the work of Gregor Mendel, which showed that the material of heredity is *particulate*, that is, not infinitely divisible but discrete, and thus that mutations can be stored in the genes without being diluted. These findings eventually caused the scientific community to revive the idea of natural selection as the mechanism for evolution, despite the fact that Darwin had abandoned the idea.

These two models of inheritance, *pangenesis*, implying the inheritance of acquired characteristics (the Lamarckian view), and *natural selection*, were opposed to one another. There were considerable data available that bore on the two models. They could easily have been compared probabilistically, as part of the scientific analysis. Such an analysis, and related analyses about the model of evolution versus the model of Biblical genesis, have not yet been carried out, although the arguments surrounding such competing models still rage. As in the case of Galileo and the heliocentric and geocentric theories, the arguments regarding natural selection and evolution go far beyond science and mathematics by pitting religious belief against the opinions of the scientific community. Although it is likely that probability would have helped somewhat both in Galileo's time and in Darwin's, the issue of convincing the public is one that pits science against religion.

Louis Pasteur has been shown within recent years to have let his strong prior beliefs influence his clinical treatments and his reports of data he falsely claimed to have acquired through experimentation. On the basis of theory and its demonstrated effectiveness in the case of the microbe for chicken cholera, Pasteur believed that atmospheric oxygen would be effective in attenuating the virulence of other microbes. He was extremely enthusiastic about this theory of the usefulness of atmospheric oxygen. Hence in reporting a highly successful public test of the effectiveness of a vaccine against anthrax, Pasteur claimed that he had prepared the vaccine using atmospheric oxygen, when in fact it had been produced using potassium bichromate. Similarly, when he believed he had hit upon a process that would produce immunity to rabies, Pasteur felt himself justified in using that vaccine on two young boys who had been bitten by a rabid dog. He claimed that he had already successfully immunized 50 dogs with this same vaccine; his recently revealed laboratory notebooks make it clear that he had barely begun trials of this vaccine and had as yet no results at all. In his work to disprove the doctrine of the spontaneous generation of life, Pasteur is seen to have suppressed the results of experiments that had what he considered "errors," that is, data that might lend support to the theory he was trying to disprove. We now know that Pasteur's experimental procedure generated insufficient heat to kill all the microorganisms originally present in his experimental material, so that the "erroneous" results do not, in fact, support the idea of spontaneous generation. But Pasteur himself did not have this explanation available. Regardless, Pasteur's sterilization process was eventually justified, and resulted in the pasteurization process that has currency today.

One of the several model comparisons that Pasteur was involved with was the one that examined the theory of spontaneous generation, and he confronted it with the theory that life can only derive from other life. He introduced the notion of sterilization (now called pasteurization) to prove his point. Would probabilistic analysis have helped his cause? Probably, but despite all his brilliance and flashes of scientific insight, Pasteur's approach to data was to suppress them if they didn't agree with his preconceptions of what they ought to be. Bayesian analyses by others of data analogous to Pasteur's would surely have helped Pasteur make his case more convincing.

Partly because Sigmund Freud used subjectivity so pervasively in his work, some observers believe that his work is not science at all. He used data based on patients' free associations and recounting of dreams to interpret their subconscious thoughts, and thus derive what he considered universal laws indicating that the sexual urge and the libido are the driving forces behind all repression and inner conflict. Observers of Freud's work have stressed that his ideas evolved in a mind suffused with a sense of personal destiny and that maintained a lively interest in the occult. His attention to mysticism perhaps stemmed from his father's Chassidic mysticism and from his own heavy use of cocaine. Because they argue that Freud never scientifically tested his claims for the causal links he proposed between sexual repression and neurosis, and between the processes of psychoanalysis and cure, some critics feel Freud's work is pure subjectivity (rather than the required combination of subjectivity and objectivity found in all good science). Freud's data are questionable in a scientific sense since he did not carry out controlled experiments and his observations were totally subjective assessments of what was happening to his patients. But that has been the nature of most of the field of psychoanalysis, although there are positive signs that the stance may be undergoing change in the appropriate direction.

Marie Curie was long sustained in her investigation of radium by her belief that the new substance was an element. But her belief that elements could not be transmuted to other elements (even through radioactive decay) kept her from being able to explain the origin of radioactivity, although she (together with her husband, Pierre) originated the insight that radiation was an atomic property. The fact that the elements with which she was working (not only radium, but polonium as well) had long half-lives made it extremely difficult for Curie to measure the loss of weight and energy due to radioactive decay. Thus, despite her toying in a 1900 publication with the idea of transmutation, Marie Curie's strong subjective prior belief in the established principles of physics, coupled with her strong belief in what her data told her, kept the next insight from her. A Bayesian model comparison of alternative theories proporting to explain radioactivity would probably have directed her towards such insights.

Albert Einstein was a *theoretical* physicist in the sense that he derived his beliefs about the physical world not only from a profound understanding of the way the physics of the world works, but also from the mathematical models that he developed of relationships that physical phenomena should obey. His equations governing the special and general theories of relativity, for example, were used to make corrections to Isaac Newton's laws of motion and to generate predictions about what might be observed in appropriate hypothetical circumstances, as when traveling at speeds near the speed of light. But he never tested any of his theories experimentally; he merely stated them subjectively and mathematically, fully expecting them to characterize correctly what would be observed in the real world. He also believed that if data were collected to test his theories, and if the data contradicted theory, he would be inclined to reject the data as erroneous, because they contradicted the theory that he simply knew a priori to be true.

Einstein was fully capable of carrying out any mathematical analysis of his choosing. So to examine competing theories of physics using probability and posterior odds ratios would have been well within his grasp. He even used probability to develop the (predictive) law of Brownian motion, one of his famous papers of 1905. But he had difficulties with collecting his own data. He was not an experimentalist. So it was not his approach to decide about any physical laws on the basis of being convinced by a preponderance of scientific data, expressed probabilistically or otherwise. Einstein just intuitively and mathematically decided how things ought to be and then enunciated what the general principles had to be. Probability considerations are very unlikely to have been much of a cogent force on his own beliefs, although they might have strengthened the scientific beliefs of others in his theories (such as in the theory of general relativity) at earlier stages in his career.

We can discern several themes in this recital of the earlier uses of informal subjectivity in science and the use of formal probability to examine competing scientific hypotheses. First, we see generalizations from data (induction), precisely as Bacon had prescribed (and contrary to Popper's views involving falsification). But sometimes, these generalizations go well beyond the data at hand and can lead either to correct inferential leaps or mistaken overgeneralizations. We have seen Harvey correctly assume that the origin of all life is in the ovum, even though he was unable to understand the significance of the ovaries in mammals. But we have also seen Aristotle overgeneralize from watching extremely light and heavy objects fall. He concluded that heavy objects tend to fall toward the earth and light ones tend to rise, and thus that the speed of fall is related to an object's weight.

Second, we have seen instances in which a scientist, overwhelmed by a theory, is apt to see what the theory prescribes (or is unlikely to see what the theory denies). Thus Margaret Mead, expecting to see a stress-free adolescence in Samoa to justify a nurture view of the nature–nurture controversy, reported seeing just that. Marie Curie, guided by a belief in the conservation of energy, failed to identify the mechanism that gives rise to radioactivity. Albert Einstein insisted that any data that might seem to contradict his theory would prime facie be mistaken. Note how this statement echoes Galileo's expression of awe at learning that the adherents of the Coperinican (helio-centric) system were able to ignore the evidence of their own senses (that the Sun appears to rotate around Earth) to conclude that Earth revolves around the Sun.

Third, we have seen instances in which a scientist becomes sufficiently convinced of the correctness of the line of investigation being followed that a singleminded stubbornness develops. We have seen Marie Curie insisting that her newly discovered substance, radium, is an element, and we see Michael Faraday laboring for a decade to demonstrate the induction of electricity from magnetism.

Finally, we have seen instances of subjectivity that fall on the border of outright fraud—and instances that fall squarely within the domain of the fraudulent. Johannes Kepler, Gregor Mendel, Robert Millikan, and Cyril Burt reported data that were too good to be true but that in each case fit the theory that the scientist was trying to prove or to demonstrate (see Chapter 3). The data doctoring in three of these cases was in the service of a theory now believed to be true; in Burt's case the jury is still out. Sir Isaac Newton altered data between editions of his masterpiece, *The Principia*, in order that his theoretical contentions appear in the best possible light. Both Antoine Lavoisier and Louis Pasteur claimed to have carried out experiments that were either incomplete or even not yet begun at the time they published the supposed results. We do not condone their bending, or even breaking, the accepted rules of honesty applicable no less in the practice of science than more broadly in normal human interaction. Nevertheless, we must note that Newton, Lavoisier, and Pasteur were most often correct in their scientific conclusions and that they enormously advanced the cause of science with their thinking.

Throughout this volume we have mentioned that even eminent scientists may have overstepped the bounds of acceptable scientific methodology at times. But we have had no intention of trying to tarnish the eminence of the scientists whose works we have described and whose subjectivity we have documented. Indeed, the modern world is indebted to these scientists for many benefits, ranging from the comforts and safety of everyday life to sophisticated understanding of the physical and mental universes. But a consideration of such infractions of the rules, along with creative subjectivity and prediction on the basis of understanding of underlying theory, demonstrates in bold relief the processes of subjectivity that we believe to occur in much less obvious form in the day-to-day practice of science.

Subjectivity occurs, and indeed should occur, in the work of scientists; it is not just a factor that plays a minor role that we need to ignore as a flaw that sometimes creeps into otherwise "objective" scientific analysis. Total objectivity in science is a myth. Good science inevitably involves a mixture of subjective and objective parts. In Chapter 5 we'll see how the subjective and objective parts can be integrated in a natural way.

CHAPTER 5

Subjectivity in Science in Modern Times: The Bayesian Approach

The Reverend Thomas Bayes

5.1 INTRODUCTION

We saw in Chapters 3 and 4 that down through the ages, the world's most famous scientists have brought their own personal beliefs, insights, and intuition (i.e., their subjectivity) into their scientific research methodology in a variety of ways, often unacknowledged and informal. Their informal subjectivity involved educated or informed ideas, intuition, strong beliefs, scientific understanding of underlying principles, and so on, and such pre-existing ideas were generally based on earlier observations made by themselves or by others. The scientific conclusions reached by scientists generally result from a mixture of, on the one hand, belief and understanding held prior to carrying out some experiment and collecting data, together with, on the other hand, largely objective analysis of experimental data. (Instances of purely theoretical work with no data to back up the theory are special cases of such a mixture, and occur only rarely, as in the case of scientists such as Albert Einstein.)

We pointed out above that the subjective approach in the past has been "informal." We mean that for the scientists we discussed in Chapters 3 and 4, as well as for many others, it is generally difficult to state which portions of their postexperimental conclusions and beliefs about some phenomenon are attributable to subjectivity and which to largely objective observation. In modern times (especially since the second half of the twentieth century), many scientists have continued to bring subjective knowledge about the phenomena they are researching to bear on their analyses in informal ways. But others have begun to take advantage of the benefits of a more formal approach, one that uses the methods of *Bayesian statistical analysis*. In the discussion below, we will see how formal subjectivity is used and how it is incorporated in Bayesian statistical analysis of scientific data. In this context, the term *subjectivity* refers to beliefs, information, or knowledge held by the scientist prior to taking new data. In this chapter we explain *some* of the ways in which subjectivity is being used in modern times, by both scientists and laypeople. This subjectivity is used in decision making in everyday life and at all levels of policymaking in business, industry, and government. It is even used sometimes in cases when no new data are collected at all.

Some articles that solve real problems using the Bayesian approach in many diverse disciplines are cited in the Appendix. A recent article discusses the importance of, and the rate of the shift to, the Bayesian paradigm that the scientific establishment is currently undergoing (Malakoff, 1999). Malakoff pointed out, for example, that the number of scientific Bayesian papers that have been published each year has steadily increased since 1991, and reached about 1400 in 1999, a trend that he speculated might be called "a Bayesian publishing boom." He also elaborated on the diversity of fields of application where Bayesian statistical methods have been applied: "from astrophysics to genomics and in real world applications such as testing new drugs and setting catch limits for fish. The long-dead minister is also weighing in on law suits and public policy decisions and is even making an appearance in consumer products" (p. 1460).

5.2 ORIGINS OF PROBABILITY AND STATISTICAL INFERENCE IN SCIENCE

Use of probabilistic and statistical methods in science evolved from the earliest days of mercantile trade, when people began to calculate the appropriate cost of insuring themselves to cover the economic risks they were considering taking in business ventures. (How does the chance of encountering some undesirable event relate to the cost that would be so incurred?) So started the field of insurance. Later, in the seventeenth century, gamblers approached mathematicians with questions about the chances of some random event occurring. The events they were interested in were those they were considering betting on, such as the chance that double sixes would come up in throwing a pair of fair (well-balanced) dice cast in a gambling game. The mathematicians then worked out the probability laws required to answer such questions. Mathematical probabilities are calculated by enumerating the number of possible ways an event such as double sixes can happen (it can happen in only one possible way) and expressing that as a fraction of the total number of possible outcomes of the dice. A normal pair of dice can fall in 36 possible ways, hence the probability of double sixes for a pair of well-balanced dice is 1/36. But such purely enumerative calculations are useful only if all outcomes are equally likely and only if the number of possible outcomes is finite, so that we can carry out the enumeration process. But what if all possible outcomes are not equally likely? With specially designed dice certain outcomes are more likely than others. For a loaded die, one side is more likely to fall uppermost than are others. For another example, because it is so difficult to produce a high quality computer chip, in the manufacturing process, a defective computer chip is currently a much more likely outcome than an acceptable, nondefective, chip.

Suppose that one or both of a pair of dice is loaded (weighted so that one number is more likely to appear than others). Then simple counting rules cannot be used to define probabilities. Suppose, for example, that we know that one die is loaded so that "six" is more likely to come up than are the other sides. Thus, if both dice are tossed, we know that double sixes is more probable as an outcome than the 1 time in 36 that we calculated for a pair of well-balanced dice. But how much more probable? One way that we can find out is by tossing the dice repeatedly, counting the number of times we get double sixes, and expressing that count as a proportion of the number of tosses. This is called the *long-range relative frequency* definition of probability. It is clear that when the dice are not loaded, the probability of double sixes we get by the method of counting sides will be duplicated by a long-range relative frequency, also equal to 1/36 (assuming that the dice are cast a very large number of times). The long-range relative frequency definition of probability relates the important construct of mathematical probability to something that can be observed empirically, and thereby marries it to empirical science.

For all its advantages, the long-range relative frequency definition of probability made the term (and the laws for manipulating probabilities) applicable only to situations that could, conceptually at least, be repeated many times. But there are many other sorts of situations, ones that are unique in that they cannot, even conceptually, be repeated many times, to which we often want to apply the notion of probability. For example, we may be interested in the probability that this scientific theory predicts future observations well (we discuss this important matter more extensively below), the probability that it will rain during the next hour, or the probability that our candidate will win in the upcoming election. These events will occur only once. It became clear that the definition of probability needed to be expanded to one that that would cover all such situations and would include the earlier definitions as special cases.

Eventually, probability came to be thought of as an individual's personal degree of belief about some proposition or about any quantity unknown to the person making the probability statement. The individual's personal degree of belief about an event, or *personal probability*, is often called a *subjective probability*. The term *subjective* used here implies that this construct is based on a belief specific to an individual. The other definitions of probability given above are special cases of subjective probability. The belief could be that the mathematical probability (an enumerative proportion) is what is called for in a given instance. Or it might be that the long-range relative frequency (the proportion of times some event occurs within many trials) is what is called for in the belief system of the person making the probability statement. Or it could be that neither definition is appropriate, and the subjective probability statement instead implies a belief of the individual based on his or her scientific knowledge and deep understanding of some underlying biological, physical, or social phenomenon.

For example, consider the proposition "Andrew Jackson was the eighth president of the United States." This proposition is either true or false. You can look it up in a reference book if you're not certain about it. But you personally may not know whether the proposition is true or false, so for you it is uncertain; but you have a *degree of belief* about its correctness. You may feel there's a 50:50 chance (50% probability) that it's correct, or you may feel, for example, that it's almost certainly true, because of other information you have in your head about other presidents, about events of that period of history, or whatever. You might actually have *any* degree of belief (probability) whatever about this proposition. The point is that it is *your* personal probability for the truth of this proposition, and you might even be willing to bet on it. For example, you may feel there's a 90% chance that the proposition is true (equivalently, that there's just a small degree of uncertainty in your mind about the truthfulness of this statement). It might also mean that you would be happy to make a wager with me by which I give you a dollar if it turns out that the proposition is true, and you give me $9 if it turns out that the proposition is false. Equivalently, if you believe that there's a 90% chance that the proposition is true, you believe that the chance that the proposition is true is the same as the chance of randomly drawing a white ball out of an urn that contains 90 white balls and 10 black balls, all randomly mixed up. (By the way, Andrew Jackson was actually the seventh president of the United States, serving 1829–1837.)

We could make an analogous argument about any quantity unknown to you, such as the height of the next person you might see accidentally in the street or whether a particular scientific theory is true. Your personal belief about the height of the person in the street would most certainly not be that the height is greater than 10 feet, or as small as 5 inches; you know in advance that people are not that tall or small. So you could say that your subjective or personal probability that the height of the person is less than 5 inches or greater than 10 feet is zero. But your subjective probability might be that there's a 50% chance that the person's height is within an inch or so of 5 feet 8 inches, that there's a 25% chance that the person's height is less than 5 feet 4 inches, and a 15% chance that the person's height will exceed 6 feet. These subjective probabilities are based on your own previous observations, experiences, and understandings about human growth and development. Indeed, if you thought that the next person you will see is, in some sense, randomly chosen from the population, you might use all the published information about the distribution of heights in the population to form your personal probability.

You might have a belief, before conducting any experiments, that the probability that a particular theory is true is 50%. You might then collect observational data that add more evidence that the theory is true, to the extent that you are willing to say that your probability that the theory is true has increased to 75%. In fact, a scientist is never certain (100% probability) that a theory is true; he or she only knows that this theory predicts future observations well and that no contradictions to the theory have yet been observed.

By comparison with probability, *statistical inference* really began only in the latter portion of the nineteenth century and began to flourish as a discipline starting only in the 1930s. [For a discussion of the development of statistical inference see, for example, Stigler (1986).] Statistical inference involves scientists collecting observations about some phenomenon, and from those observations, trying to infer or induce some general underlying principles. Those principles are then used to predict outcomes of future experiments that involve that phenomenon (see Section 5.6).

5.3   BAYESIAN STATISTICAL INFERENCE: PRIOR AND POSTERIOR BELIEF

In 1763 an important scientific paper was published in England, authored by a Reformist Presbyterian minister by the name of Thomas Bayes (see Bayes, 1763). An implication of this work was a method for making statistical inferences that builds on earlier understanding of a phenomenon, and that *formally* combines that earlier understanding with currently measured data in a way that updates the scientific belief (subjective probability) of the experimenter. The earlier understanding and experience is called the *prior belief* (subjective knowledge in the form of belief or understanding held prior to observing the current set of data). The new belief that results from updating the prior belief by combining it with the experimental data is called the *posterior belief* (the belief held after having taken the current data and having examined those data in light of how well they conform with preconceived notions). This inferential updating process is eponymously called *Bayesian inference*. The inferential process suggested by Bayes shows us that to find your subjective probability for some event, proposition, or unknown quantity, you need to multiply your prior beliefs about the event by an appropriate summary of the observational data. Thus, Bayes pointed out to us that all formal scientific inference inherently involves two parts, a part that depends on the subjective information and understanding that you have prior to carrying out an experiment (prior knowledge), and a part that depends on scientific observation and experiment (a summary of the observational data).

A large international school of scientists preceded, supported, expanded, and developed Bayesian thinking about science. These included such famous scientists as James Bernoulli in 1713 (even before the paper by Bayes was published), Pierre Simon de Laplace in 1774, and many twentieth century scientists such as Bruno de Finetti, Harold Jeffreys, L. J. Savage, Dennis V. Lindley, and Arnold Zellner. Bayesian methodology was the method of statistical inference generally used from the time of Bayes until the early part of the twentieth century, when Sir Ronald A. Fisher and others introduced the *frequentist approach* to statistical inference. To many, it has become increasingly clear that the frequentist approach is fraught with technical problems and inconsistencies (see, e.g., Berger and Berry, 1988; Howson and Urbach, 1989; Matthews, 1998a,b). As a consequence, today, scientists schooled in the Bayesian approach to scientific inference have been departing from the frequentist approach and returning to the Bayesian approach. Many scientists now believe that a paradigm shift in the sense of Kuhn (1962) has been taking place in the way that scientific inference is, and will be, carried out. Many scientists now recognize the advantages of bringing prior beliefs into the inference process in a formal way from the start. The alternative is attempting to achieve total objectivity, or pretending to have done so, even though prior information is often brought to bear on the problem anyway, in surreptitious or even unconscious ways. Subjectivity may enter the scientific process surreptitiously in the form of seemingly arbitrarily imposed constraints, in the introduction of initial and boundary conditions, in the arbitrary levels of what should be called a statistically significant result, in the deemphasizing of certain data points which represent suspicious observations (this is what was done in the case of Robert Millikan, as discussed in Section 3.4), and so on.

The Bayesian result is that the degree of a person's belief about some unknown entity, once something about it has been observed, is determined as the product of two types of information. One type of information characterizes the data that are observed; this is usually thought of as the "objective" portion of posterior belief, since it involves the collection of data, and data are generally referred to as "objectively determined." (We recognize that we do not really mean that data are necessarily objective unless we assume that there were no subjective influences surrounding the data collected. See the discussion in Chapter 1 about the perceptual influences on data collection as well as the subjective influences on data interpretation exemplified by the rate-of-defectives example.) This so-called "objective" information is summarized in what is sometimes called the *likelihood function*. It summarizes the observational data from an experiment in a probability form.

The other type of information used in Bayesian analysis is the person's degree of belief about the unknown entity held prior to observing data obtained from the current experiment. (The prior belief might be based on theory, or implications of some mathematical formulation of the theory, or it might be based on a combination of theory and empirical data obtained in earlier experiments.) Thus, this prior belief may be based, at least in part, on things that were observed or learned about this unknown quantity or theory prior to this most recent measurement. The product of the probability of the data (based on a given model for the theory) and the prior (subjective) probability of the theory determines the posterior (subjective) probability. Bayesian analysis of scientific data formally combines these two types of information.

We saw in the rate-of-defectives example in Chapter 1 that several scientists examining the same set of data from an experiment came up with different interpretations. This phenomenon is not unusual in science (or in everyday interaction; see the film *Rashomon*). When several scientists interpret the same set of data, they rarely have *exactly* the same interpretations. Almost invariably, their own prior beliefs about the underlying phenomenon enter their thinking, as do their individual understandings of how meaningful each data point is. Their conclusions regarding the extent to which the data support the hypothesis will generally reflect a mixture of their prior degree of belief about the hypothesis they are studying and the data observed.

Whether formal Bayesian inference is actually used in dealing with the data in an experiment, or whether other, non-Bayesian methods are used, prior belief is generally used in one way or another by all good scientists in a natural, sometimes quite informal, way. Science cannot, and should not, be totally objective, but should and does involve a mixture of both subjective and objective procedures, with the one type of procedure feeding back on the other. As the data show the need for modification of the hypothesis, ideally, a new hypothesis is entertained, a new experiment is designed, and new data are taken. What had been the posterior belief in the earlier experiment becomes the prior belief in the new experiment, because the result of the last experiment is now the best understanding the scientist has of what result to expect in a new experiment. To study the future, scientists must learn from the past; and it is important—indeed inevitable—that the learning process be partly subjective in that the scientist must continually revise his or her information about a phenomenon in the light of new information.

The best information scientists (and the lay public) have about a particular phenomenon is the collection of beliefs and results of experiments carried out by one or more particular scientists who have been investigating that phenomenon. Then other scientists repeat those experiments for verification; they may find that their results approximate those of the original scientists who carried out the seminal investigations. Then the posterior probabilities of this collection of scientists about this phenomenon become those of the general public.

5.4 MATHEMATICAL FORMALISM OF THE BAYESIAN APPROACH

(The less mathematically inclined reader may wish to skip this section.) Although it is not necessary to read this section for general understanding of the formal subjective approach, for those who are comfortable with mathematical formulations, we offer a brief mathematical explanation of the very simple Bayesian formula that underlies all of Bayesian statistical science, the modern approach to subjectivity in science. We assume that there is some underlying experiment of interest and that the experiment has more than one possible outcome; that is, the outcome of the experiment is uncertain and unknown until the experiment is actually performed, and there are several possible outcomes. (The number of possible outcomes might be two, or three, or four, or more, or even an uncountably infinite number, such as any number in an interval.)

Suppose we call two of the possible outcomes that can occur in an experiment, event A and event B. For simplicity of illustration, we assume that as the outcome of the experiment, event A may or may not occur, event B may or may not occur, both may occur, or neither may occur. We use the traditional symbols and formalism that "$P\{\text{event}\}$" denotes the degree of belief, probability, or chance, that a particular event will occur when a random mechanism generates an event. The probability is calculated in advance of the experiment being performed. Once the experiment has been carried out, there is no longer any uncertainty about the outcome; the outcome is known.

Bayesian statistical analysis depends fundamentally on a relationship derivable from the axioms of probability theory that enables us to convert the chance that an event A will happen, conditional on our having information about event B into the chance that event B will occur once we know that event A has already occurred. Then, using this symbolism, Bayes' formula asserts that

$P\{(\text{event } B \text{ will occur}) \text{ given } (\text{event } A \text{ has occurred})\}$

$$= P\{B \text{ given } A\} = \frac{P\{A \text{ given } B\} \times P\{B\}}{P\{A \text{ given } B\} \times P\{B\} + P\{A \text{ given not } B\} \times P\{\text{not } B\}}.$$

In this formula, $P\{A \text{ given } B\}$ is generally referred to as the *model*; it depends on the observational data, as we will see. $P\{B\}$ is generally referred to as the *prior probability* (since this probability depends only on information known prior to observing the data). $P\{B \text{ given } A\}$ is referred to as the *updated* or *posterior probability* of B since it is conditioned on having information about event A. (The prior probability of B was not so conditioned.)

5.4.1   Example 1: A Bayesian Inference Problem in Medical Diagnosis

Suppose that you go to the doctor for a regular medical checkup. In the course of your examination, the doctor suggests a laboratory test for a serious illness, such as a particular form of cancer. For the purpose of illustrating the formal subjective (Bayesian) approach to scientific inference, we will keep the example very simple by assuming (unrealistically) that the doctor has available only the one test. We also assume that she has not discovered any reason to believe, a priori, that you have the disease—you have no history of the disease, it's not in the medical background you've told her about, and she doesn't see any particular signs of the disease otherwise. She just believes everyone should be tested for this disease as a matter of routine.

Suppose that the medical test is a very sensitive one in that the chance that the lab test is positive when in fact you don't have the disease (*false positive*) is 3%, and the chance that the lab test is negative when in fact you do have the disease (*false negative*) is 1%. (The false negative kind of error is, of course, of greater concern to you.) These error rates for the test reflect what happens on the average. In other words, for example, the test doesn't necessarily yield three false positive results for every 100 times the test is administered on randomly selected patients. Sometimes, it won't give any false results in thousands of tests, and other times it may give 10 or more false results in 100 tests. But on average, the false positive rate is 3%; similar possibilities hold for the false negative rate.

Suppose also that the incidence of the disease in the population at large is 1 in a million (you adopt this value as the prior probability that you have the disease, $P\{B\}$ in the notation above). You take the lab test, and to your surprise and horror, the lab result comes back positive! What should you now believe about the chances that you actually have this dreaded disease? (That is, what is the posterior probability that you have the disease? This is $P\{B \text{ given } A\}$ in the notation above.) In light of the laboratory evidence, our fictitious physician whose tools we have limited to just this one lab test (and who might be unfamiliar with probability) might just say that there's a 97% chance that you actually do have the disease, since the physician has been advised that when the lab test is positive, it is fallible only 3% of the time. In fact, a more precise and less threatening interpretation of the lab result is that if you don't have the disease, the test will, on the average 3% of the time, indicate *incorrectly* that you do have the disease. A computation (provided below) using the Bayesian paradigm would say that the chances that you actually have this form of cancer are only 0.003%, less than a mere one hundredth of 1 percent! Our fictitious physician, who is unsuspecting of her misinterpretation of the error probabilities of the test, has given you very bad, and incorrectly disheartening, advice. Of course, we also handicapped her by making only one laboratory test available to her for diagnosis.

The Bayesian paradigm asserts above that in the situation created for this medical diagnosis example, the probability that you actually have the cancer in question, taking into account that the lab test was positive, is given by

$P\{$(you have cancer) *given* (your positive lab test)$\}$
$= P\{$(cancer) *given* (+test)$\}$

$$= \frac{P\{(+\text{test})\ given\ (\text{cancer})\}P\{\text{cancer}\}}{P\{(+\text{test})\ given\ (\text{cancer})\}P\{\text{cancer}\} + P\{(+\text{test})\ given\ (\text{no cancer})\}P\{\text{no cancer}\}}.$$

In this situation, we can specify the sensitivity of the test by

$P\{(+\text{test})\ given\ (\text{no cancer})\} = $ probability of a false positive $= 0.03$;
$P\{(-\text{test})\ given\ (\text{cancer})\} = $ probability of a false negative $= 0.01$,

so

$$P\{(+\text{test})\ given\ (\text{cancer})\} = 1 - P\{(-\text{test})\ given\ (\text{cancer})\} = 1 - 0.01 = 0.99.$$

Substituting these numbers into the Bayesian formula gives

$$P\{(\text{cancer})\ given\ (+\text{test})\} = \frac{(0.99)P\{\text{cancer}\}}{(0.99)P\{\text{cancer}\} + (0.03)P\{\text{no cancer}\}}.$$

But we have supposed that the incidence of the disease in the population at large is 1 in a million. So you take

$$P(\text{cancer}) = 10^{-6} = 0.000001.$$
$$P(\text{no cancer}) = 1 - 10^{-6} = 0.999999.$$

Substituting these numbers gives approximately

$$P\{(\text{cancer})\ given\ (\text{positive lab test})\} = 0.003\%.$$

In view of this very small posterior probability, if you really believe subjectively that your chance of getting this cancer is equivalent to that of a randomly selected person in the population, it is quite unlikely that you have cancer despite the positive outcome of the test.

But suppose that when the doctor had asked you about your medical history, you didn't know that there had been a strong incidence of cancer in your family. On further investigation, suppose that you found that three out of four of your grandparents had died early of the same type of cancer that you are being tested for and that your father had been misdiagnosed when he died. In fact, your father actually died from this form of cancer. Now you have to question the figure of 1 in a million that applies to the population at large. It probably doesn't apply to *you*. The chances of getting this type of cancer may be much higher for *you* than for others, because of something related to *your* genes. So the incidence of cancer for *your subpopulation* is not 1 in a million, but perhaps it is 1 in 1000. Now, as it turns out, *your* chance of having this type of cancer given your positive test result increases to about 3.2%. The chance is still much smaller than the 97% the fictitious doctor had mistakenly conjectured, but it is larger than what it might be calculated to be without taking into account the subjective factor that relates to *you*.

We can readily carry out analogous computations using the Bayesian paradigm considering a collection of different incidences of the disease, perhaps appropriate for different subpopulations to which you might believe that you belong. We find that the probability that a hypothetical person with a positive test result actually has the disease doesn't reach even 50% until the incidence of the disease in the appropriate population is 1 in 34 (see the table "Probabilities for a Disease").

Probabilities for a Disease[a]

Incidence of Disease (Prior Probability)	Posterior Probability (%)
1 in a million	0.003
1 in 100,000	0.033
1 in 10,000	0.33
1 in 1000	3.2
1 in 100	25
1 in 34	50
1 in 10	78.6

[a] 3% probability of false positives and 1% probability of false negatives

5.4.2   Value of a Second Opinion

Now suppose that because you feel that cancer is such a serious disease, you decide to seek a second opinion about your problem. You look for another physician, perhaps an oncologist, who will suggest either that a new, independent lab test be obtained, or that a new pathologist evaluate the results of the same lab test. The purpose of either procedure is to see whether a new test or a new reading of the same lab results will change your opinion about your chance of having the disease. (Because such lab tests can be quite invasive, and even dangerous, multiple readings of the same lab results are often taken rather than additional invasive procedures being performed.) In either case, the result will be either positive again (confirmatory), or perhaps this time it will be negative.

Suppose the second lab test is positive again. How should that affect your belief about your having this cancer? Bayes' formula can be used again. Before the second test results are known, you best information about your chances of having the disease is the posterior probability that you developed on the basis of the first test results. That probability was originally 0.003%, based on using the 1 in a million incidence of the disease in the general population. But then when you determined that this particular type of cancer ran in your family, you modified your belief to 3.2%. Now this posterior probability becomes your new prior probability. So you can reapply Bayes' formula using a prior probability of 3.2% (assuming that the false positive and false negative probabilities remain the same). The resulting calculation shows that the new posterior probability that you have the cancer in the light of the second test results has risen to 52.2% (or 16 times the earlier probability). Very sensibly, you start treatment as soon as possible, and perhaps simultaneously, you seek even a third opinion.

It is important to bear in mind that although this example is illustrative of one type of use of the Bayesian approach, we assumed unrealistically that the original doctor is limited to one test. In fact, physicians generally depend for their diagnoses upon several tests and symptoms displayed by their patients. It is only after certain symptoms persist over time, and several tests of various kinds point to the same diagnosis, that meaningful conclusions can be drawn. The Bayesian approach can be used to great advantage in this context, as we have seen, but sometimes the problem is more complicated and the additional complications must be taken into account in the Bayesian analysis.

5.4.3 Example 2: Bayesian Inference in Physics

(To understand this example the reader should have some familiarity with the *binomial distribution*.)

Suppose that a scientist is interested in evaluating his posterior probability that there is a particular physical effect operating in his experiments. Denote the effect by E. Suppose he feels a priori that there is a 50:50 chance that the effect is present. To evaluate the effect, the scientist knows that he must first do some preexperiment trials. These may terminate either in his observing some minimum voltage (called a *success* because it tends to indicate that the effect is happening) or not (called a *failure* because it tends to indicate that the effect is not happening). He carries out the trials independently of one another, and the chance of a success on a single trial remains the same from one trial to the next. He carries out the trials three times and finds two successes. These are the observational data. The scientist calculates that if the effect E is actually present, the chance of a success on a single trial is 0.6, but if E is not present, the chance of a success is just 0.2.

Bayes' formula gives the posterior probability of E given the 2 successes in 3 trials as

$$P\{E \text{ given } 2 \text{ successes}\}$$
$$= \frac{P\{(2 \text{ successes}) \text{ given } (E)\}P\{E\}}{P\{(2 \text{ successes}) \text{ given } (E)\}P\{E\} + P\{(2 \text{ successes}) \text{ given } (\overline{E})\}P\{\overline{E}\}}.$$

Here $\overline{E}$ denotes the event "not E," meaning the event that the effect E is not present. We note that because of the 50:50 chance assumed in this example $P\{E\} = P\{\overline{E}\} = 0.5$. Also (using the binomial distribution), we find that $P\{(2 \text{ successes}) \text{ given } E\} = 0.432$, and $P\{(2 \text{ successes}) \text{ given } \overline{E}\} = 0.096$. Substituting into Bayes' formula gives

$$P\{E \text{ given } 2 \text{ successes}\}$$
$$= \frac{P\{(2 \text{ successes}) \text{ given } (E)\} \times (0.5)}{P\{(2 \text{ successes}) \text{ given } (E)\} \times (0.5) + P\{(2 \text{ successes}) \text{ given } (\overline{E})\} \times (0.5)}$$
$$= \frac{P\{(2 \text{ successes}) \text{ given } E\}}{P\{(2 \text{ successes}) \text{ given } E\} + P\{(2 \text{ successes}) \text{ given } \overline{E}\}}$$
$$= \frac{0.432}{0.432 + 0.096} = 0.818.$$

Summarizing,

$$P\{E \text{ given } 2 \text{ successes}\} = 0.818 = 81.8\%.$$

So the scientist's posterior probability that there is an effect E has increased from 50%, a priori, to 81.8%.

Applications of Bayesian analysis to problems in physics abound, for example, in the work of Harold Jeffreys (see Jeffreys, 1961) and Edward Jaynes (see Jaynes/Rosenkrantz, 1983). The approach was also advocated by Richard Feynman [see the book by Mathews and Walker (1964) in which the authors explain in the Preface that the book was an outgrowth of lectures by Richard Feynman at Cornell University; see pp. 361–370, where Feynman suggests that to compare contending theories one should use Bayesian methods. See also, the example in Section 5.6.2 below].

The previous two examples were applications of Bayes' formula for cases in which the data followed discrete probability distributions. (Sections 5.4.1 and 5.4.2 involved some basic discrete probabilities associated with a problem in medical diagnosis, and Section 5.4.3 involved the binomial distribution with some physical science experiments.) But Bayes' formula can readily be applied to much more sophisticated, realistic problems in science. We summarize such a problem in the following example.

5.4.4   Example 3: Bayesian Inference in Biological Science

[This example is a very brief summary of an article by Gary A. Churchill that appeared in Gatsonis, Hodges, Kass, and Singpurwalla (1995, pp. 90–138, with discussion; see the Appendix).]

The information about heredity in an organism is found in its *genes*. The genetic material is called a *DNA molecule*. The DNA molecule is composed of subunits called *bases* (nucleotides), of four types: adenine (A), cytosine (C), guanine (G), and thymine (T). Information about heredity is encoded in a DNA sequence according to the specific ordering of the bases. There are two strands of DNA (the *double helical structure* first proposed by Watson and Crick in 1953) in each gene, and they are joined together in a complementary way so that that an A on one DNA strand is always paired with a T on the opposite strand, and D is always paired with C. The complete genetic information is therefore contained in the sequence of just one strand.

The totality of DNA in an organism is called its *genome*. There has been extensive activity in the biological community in recent years to map the DNA sequence of many organisms, especially to map the human genome containing about 3 billion base pairs. There are automatic sequencing devices for accomplishing this feat, but the sequencing must still be carried out very slowly. The large number of base pairs has exacerbated the problem. The sequencing is most frequently carried out by a process called the *enzymatic method*.

At this point the scientific Bayesian analysis problem becomes quite complicated and beyond the scope of this volume. Suffice it to say that prior information is used to describe the frequencies and types of errors that arise in evaluating the sequences in separate portions of the DNA sequence (DNA fragments). The prior distribution often used in such an analysis is not discrete, but continuous, and such analysis, although too advanced to be discussed here, is treated in books on Bayesian statistical science (see Appendix A). The posterior distribution is also continuous, and to evaluate its characteristics, sophisticated, computer-intensive, statistical methods of sampling from the posterior distribution must be used (called Markov Chain Monte Carlo sampling).

One of the most important challenges to data analysis in modern statistical genetics and molecular biology is the problem of how to analyze gene expression data in which the number of variables or dimensions, "p", greatly exceeds the number of replicates, "n" (that is, we have p much greater than n). Such data frequently are generated by a piece of equipment called a microarrayer. The data analysis problem is opposite to the usual problem in statistical data analysis in which typically p is much less than n. In the microarrayer context, we might have p = 10,000, for example, while n = 10. In such contexts, to reduce the problem to one where ordinary statistical methods can be applied, it is necessary either to increase the number of replicates (sample size) or to reduce the number of dimensions, or both. The Bayesian approach may offer the best solution to such problems since in Bayesian analysis, as is easy to show, bringing prior information to bear on a model is equivalent to adding more replicates, or increasing the effective sample size.

5.5 USING PROBABILITY TO COMPARE SCIENTIFIC THEORIES

It is important for scientists to be able to demonstrate that their theories and experimental results improve in some way on the knowledge that was available prior to their work. Often, there are competing theories that are believed to explain the same phenomenon. How can scientists decide which of several proffered theories is best? Various proposals have been made for resolving this important issue. We suggest here that the ideal way to compare competing theories is by means of probability.

Our studies of the lives of scientists in Chapters 3 and 4 show that the beliefs of knowledgeable scientists about various phenomena, beliefs that are usually based, in part, on empirical data, are really the best thinking that we have available in our attempts to understand biological, physical, and social phenomena. For us, probabilities express our personal beliefs about an event or proposition. But when empirical data are available that bear on the event or proposition, these empirical data are built into the probability belief of the scientist doing the analysis, either informally by educated judgment, or formally via Bayesian methods as part of a "prior belief" (prior to taking observations).

For example, to say that you have a subjective probability (degree of belief) of 70% for the chance that some physical theory or proposition is true is equivalent to saying that your (subjective) probability for that event or proposition is the same as the chance that you will draw a white ball randomly from an urn that contains 70 white balls and 30 nonwhite balls, all well mixed up in the urn. If, instead, you do not believe that the balls were really well mixed up in the urn, but rather that the white ones were added only after the others were already put into the urn with little further mixing, so that the white ones were on top and most readily drawn, your belief that a white ball will be drawn would be greater than 70%. In that case, your subjective probability might be, perhaps, 85% (reflecting not an enumeration of the fraction of white balls in the urn, but rather, an understanding of the fact that white balls were much more likely to be drawn than their number in the urn would suggest). Thus, as Savage (1954) pointed out (and as we have mentioned earlier in this chapter), "mathematical" or "objective" probability is a special case of subjective probability.

Good (1950, 1965, 1983) proposed that to compare scientific theories, scientists should examine the *weight of the evidence* favoring each of them, and he has shown that this concept is well defined in terms of a conditioning on prior information. In the same spirit, Jeffreys (1961) developed an axiomatically based theory of subjective probability and applied it directly to problems in modern physics. In the Jeffreys approach, one can readily compare theories by examining their posterior probabilities to see which one has the greatest posterior (subjective) probability. The posterior probability of the theory being correct is the probability obtained after combining the experimental data with the belief held prior to collecting the experimental data, via Bayes' formula (used, for example, in the medical diagnosis example above).

We saw through the studies of scientists in Chapters 3 and 4 how varied scientific methodology has been over the centuries and how strong personal belief in scientific theories has been the general rule. We proceed in the spirit of Howson and Urbach (1989, p. 11), who state:

> It was the ambition of Popper, Lakatos, Fisher, Neyman and Pearson, and others of their schools to develop this idea of a standard of scientific merit which is both objective and compelling, and yet non-probabilistic. And it is fair to say that their theories, especially those connected with significance testing and estimation, which comprise the bulk of so-called classical methods of statistical inference, have achieved pre-eminence in the field. The procedures they recommended for the design of experiments and the analysis of data have become the standards of correctness with many scientists. . . . We [Howson and Urbach] shall show that these classical methods are really quite unsuccessful, despite their influence among philosophers and scientists, and that their pre-eminence is undeserved. Indeed we [Howson and Urbach] shall argue that the idea of total objectivity is unattainable and that classical methods which pose as guardians of that ideal, in fact violate it at every turn; virtually none of those methods can be applied without a generous helping of personal judgment and arbitrary assumption.

In this very pithy statement, Howson and Urbach point out that some philosophers, scientists, and statisticians of the first half of the twentieth century (see, e.g., Popper, 1959a,b, 1963; Lakatos, 1968, 1970, 1974, 1978; Fisher, 1922, 1935, 1947, 1956, 1970; Neyman and Pearson, 1928, 1933) developed an approach to statistical analysis of scientific data that was not based on probability. That approach, which attempted to be "objective" in its interpretation of data, is the *classical* or *frequency approach*. The frequency approach was readily adopted by the scientific community at large despite its many failings, and inconsistencies, and was (and is currently) used to estimate uncertain scientific quantities and to test hypotheses about scientific theories. This is a fundamentally *Popperian approach* (of *falsification*, or setting up a strawman hypothesis believed to be false and then showing that it is indeed false) to comparing scientific theories. It was adopted by Neyman and Pearson in statistical science and is what is often referred to as *hypothesis testing* and its concomitant, *significance testing*. These are terms used to refer to the classical method, that purports to determine whether some unknown, underlying, scientific quantity that was estimated is likely to have been generated at random from the noise or measurement error in an experiment, or is really a scientifically meaningful quantity. But these approaches are notorious for involving technical inconsistencies and inadequacies that arise in carrying out scientific inference (see, e.g., Basu, 1988; Press, 1989, Sec. 2.3.2, p. 32).

We do not view the issue of the injection of a scientist's educated belief into the scientific inference process as a problem. On the contrary, we accept it as an inescapable part of the process of science, and we point out how to take advantage of the personal beliefs and understanding that scientists have of phenomena of interest in the world to update their understanding of the limited amount of data they may have gathered about some matter. The Bayesian approach to analysis of scientific data provides scientists with a very natural and logical separation between the views held before taking data and the so-called "objective" views (views held after taking observational data). It sees both views as contributions to knowledge and provides a way of combining them.

A useful way to compare scientific theories using the Bayesian paradigm is to compare their "predictive distributions" for new observations. This comparison involves going from subjective understanding of a phenomenon in nature to the development of a theory or model that will predict new observations. We have seen that partial subjectivity in their scientific methodology is characteristic of scientists. We believe that the main task of science is to predict. Expressing the results of a scientific investigation in a predictive probability distribution combines these principles.

5.6 PREDICTIVE PROBABILITY DISTRIBUTIONS

Frequently, scientists test a theory by using a mathematical formulation of the theory (called the *model*) and then predicting the value of future observations from the model. Of course, it is rare that the observed values of the future observations are exactly the same as the values predicted by the mathematical model of the theory. Failure of exact prediction arises for at least two reasons. First, nature is generally much more complex than any simple mathematical formulation is able to capture. Second, all future observations must be measured, and measurements always have error associated with them (called *measurement error*) attributable to inaccuracies of various kinds in every device used for measurement. The discrepancy between predicted and observed data values is frequently referred to as *prediction error*. With a "good" mathematical formulation of a theory, most nuances of the theory will be included within the mathematical model. In a *good experiment*, measurement error is very small. So under good conditions, prediction error can be held to a very small value.

The quality of any scientific theory is measured by how well the theory predicts future observations. If the theory predicts poorly, it doesn't matter how well it performs in terms of fitting all previous data; it is not an acceptable theory. Sometimes the theory predicts well but performs poorly in fitting previous data well. Such performance is preferable to poor prediction, although ideally, the theory would perform well at both prediction and fitting previous data.

Once the current experiment has been performed, the scientist must evaluate the prediction error. Is the prediction error too large for the theory to be acceptable? How large is too large? These questions involve statistical inference. Mere calculation of the frequently used *root-mean-squared error* of the prediction (which merely looks at how far a particular prediction is from truth) is often inadequate for meaningful evaluation of the quality of the prediction. Alternatively, we provide below a brief summary of how to evaluate the predictive distribution of the experimental data. For illustrative purposes, and to keep the discussion of prediction at a simple mathematical level, we consider only a very simple case. This case involves prediction of a *discrete* random quantity whose probability distribution is that of the number of "successes" in independent trials of an experiment. We refer the interested reader through citations of other literature to more detailed and advanced mathematical treatments of prediction in the case of *continuous* quantities.

Suppose that the scientist is planning to carry out an experiment associated with a particular physical phenomenon. In connection with this experiment, the scientist wants to predict the value expected to be observed in the experiment. Call this as-yet-unobserved value, Y. Suppose further that there are just two competing theories for generating Y; call them theory A and theory B. (In a more general case, there might be more than two theories to compare.) Finally, suppose that last year the scientist (or some other scientist) carried out a similar experiment and observed the outcome of that similar experiment to be X. Now, based on all the information the scientist has about this phenomenon, the aim is to predict Y. Moreover, the scientist wants more than just the prediction of a single value; the aim is to find the probabilities that the future observation is likely to fall into particular ranges. We next resort to a mathematical explanation and an example for those readers who prefer that type of approach.

5.6.1   Probability Formulation of Predictive Distribution

(The less mathematically inclined reader may wish to skip this section.)

Using a simple law of probability, it can be shown that the *predictive probability* of the future observation Y, *given* the earlier observed result X, is a weighted average of the predicted values of Y assuming that theory A holds, and the predicted values of Y given that theory B holds. We have assumed for simplicity of explication that only theory A or theory B can hold. (This is the simplest case; more interesting cases occur when more than two theories have been proposed.)

Calculating a weighted average is like calculating the value or cost of a market basket of food (in this simple example, there are only two food items in the market basket). Using a vertical line (|) to stand for *given* in the notation used above, the weighted average for the predictive probability distribution is given by

$$P\{Y|X\}$$
$$= P\{Y|\text{theory } A\} \times P\{\text{theory } A|X\} + P\{Y|\text{theory } B\} \times P\{\text{theory } B|X\}.$$

Note that the weights here ($P\{\text{theory } A|X\}$ and $P\{\text{theory } B|X\}$) are the posterior probabilities (updated degrees of belief of the experimenter) obtained from Bayes' formula (following the previous experiment) of theories A and B, given that the result of that experiment was the data X.

5.6.2 Example: Comparing Theories Using Binomial Distribution

Suppose it has already been found that the posterior probability of theory A given the earlier observation X is given by $P\{(\text{theory } A) \ given \ X\} = 0.2$, and the posterior probability of theory B given the earlier observation X is $P\{(\text{theory } B) \ given \ X\} = 0.8$. These are the weights that will be used in the weighted average. Then

$$P\{Y \ given \ X\} = P\{Y \ given \ (\text{theory } A)\} \times (0.2) + P\{Y \ given \ (\text{theory } B)\} \times (0.8).$$

Now suppose that a particular experiment can result in only one of two ways: Either a specific effect is found (we call this outcome a success) or the specific effect is not found (we call this outcome a failure). The experiment might be repeated independently many times. Let Y denote the number of successes in (for simplicity) 2 independent trials of the experiment, and assume that the probability p of success in both trials is the same: namely, $p = 0.1$ if theory A is true and $p = 0.5$ if theory B is true. Then it is well known in statistical science that (because this hypothetical situation is characterized by a binomial distribution)

$$P\{Y \ given \ (\text{theory } A)\} = P\{Y \ given \ (p = 0.1)\} = \frac{2!}{Y!(2-Y)!}(0.1)^{Y}(0.9)^{2-Y},$$

and

$$P\{Y \ given \ (\text{theory } B)\} = P\{Y \ given \ (p = 0.5)\} = \frac{2!}{Y!(2-Y)!}(0.5)^{2}.$$

Note that $Y!$ means $(Y) \times (Y-1) \times \cdots \times (2) \times (1)$. Combining terms gives

$$P\{Y \text{ given } X\} = \frac{2!}{Y!(2-Y)!}(0.1)^Y(0.9)^{2-Y}(0.2) + \frac{2!}{Y!(2-Y)!}(0.5)^2(0.8).$$

This last expression, the predictive probability for the unknown Y (conditional on X), depends only on Y (once X is observed). Since there are assumed to be just two trials, Y, the number of successes observed can be 0, 1, or 2 (0 successes, 1 success, or 2 successes).

We can evaluate the predictive probabilities for these three cases by substituting $Y = 0$, $Y = 1$, and $Y = 2$ in the last formula. They are given by

$$P\{(Y = 0) \text{ given } X\} = 0.362;$$

$$P\{(Y = 1) \text{ given } X\} = 0.436;$$

$$P\{(Y = 2) \text{ given } X\} = 0.202.$$

Note that these predictive probabilities sum to 1, as indeed they must if they are to represent a (discrete) probability distribution.

It is now clear that although the experiment has not yet been carried out, if it were to be carried out, the most likely value of Y is $Y = 1$ (i.e., the most likely number of successes is 1) since its predictive probability (0.436) is greater than the predictive probabilities for the other possible values (0.362 or 0.202). Moreover,

$$P\{(Y \le 1) \text{ given } X\} = P\{(Y = 0) \text{ given } X\} + P\{(Y = 1) \text{ given } X\}$$
$$= 0.798 \simeq 80\%.$$

Also,

$$P\{(Y \ge 1) \text{ given } X\} = P\{(Y = 1) \text{ given } X\} + P\{(Y = 2) \text{ given } X\}$$
$$= 0.638 \simeq 64\%.$$

So there is an 80% chance that there will be zero successes or one success, while there is only a 64% chance that there will be one or two successes. But most likely, there will be just one success.

5.7 SUMMARY

A guide to some books on the theory and practice of Bayesian methodology is given at the end of this chapter. A list of some published applications of Bayesian methodology in a wide variety of scientific and professional fields is given in the Appendix.

In summary, as we have seen in this book, scientists have always used various subjective approaches in their research work. We believe that scientists will continue to incorporate subjective procedures into their analyses, but they will increasingly adopt the Bayesian formalism, using it for experimental design, for estimation, for hypothesis testing, for model checking, for averaging over many possible models, and for prediction of future events. The subjectivity of scientists has not only been an inherent factor in many discoveries, it is ingrained, and here to stay, and science is the better for it because its use permits scientists to achieve better, more accurate, and more predictable results.

5.8 ANNOTATED GUIDE TO SOME LITERATURE ON METHODS FOR BAYESIAN ANALYSIS

1. For a few *introductory treatments* of the theory and methods of Bayesian inference see, for example, Antleman (1997), Berry (1996), Blackwell (1969), Howson and Urbach (1989), Lee (1997), Schmitt (1969), and Winkler (1972).

2. A few *more advanced treatments* may be found in Berger (1985), Bernardo and Smith (1994), Carlin and Louis (1998), Dey et al. (1998), Florens et al. (1990), Geisser (1993), Gelman et al. (1995), Hartigan (1983), Kadane et al. (1999), Lad (1996), Leonard and Hsu (1999), Lindley (1965, 1972), O'Hagen (1994), Press (1989), Raiffa and Schlaifer (1961), Viertl (1987), and Zellner (1971). Many others are listed in the list of references.

3. For a *history of Bayesian inference* in science up to the beginning of the twentieth century, see Dale (1991).

4. Treatments of Bayesian *computation methods*, including *Gibbs Sampling/Markov Chain Monte Carlo* procedures, may be found in Albert (1996), Gamerman (1997), Gilks et al. (1996), Mockus et al. (1997), and Tanner (1996).

5. Bayesian *time-series analysis* is treated in Broemeling (1985), Muller and Vidakovic (1999, using wavelets), and West and Harrison (1997).

6. *Neural networks* are treated from a Bayesian point of view by Neal (1996).

7. Bayesian methods applied to *sample survey problems* are treated in Meeden and Ghosh (1997).

REFERENCES

Albert, James H. (1996). *Bayesian Computation Using Minitab*. Belmont, CA: Wadsworth.

Antleman, Gordon (1997). *Elementary Bayesian Statistics*. Albert Madansky and Robert E. McCulloch (Eds.). Cheltenham, UK: Edward Elgar.

Basu, D. (1988). *Statistical Information and Likelihood: A Collection of Critical Essays*, Vol. 45 of Lecture Notes in Statistics. New York: Springer-Verlag.

Bayes, Thomas (1763). "An Essay Towards Solving a Problem in the Doctrine of Chances," *Philosophical Transactions of the Royal Society of London*, **53**:370–418.

Berger, James O. (1985). *Statistical Decision Theory and Bayesian Analysis*, 2nd ed. New York: Springer-Verlag.

Berger, James O., and Donald A. Berry (1988). "Statistical Analysis and the Illusion of Objectivity," *American Scientist*, **76**:159–165, March–April.

Bernardo, José M., and Adrian F. M. Smith (1994). *Bayesian Theory*. New York: Wiley.

Bernoulli, James (1713). *Ars Conjectandi* (The Art of Conjecturing), Book 4. Basileae: Impensis Thurnisiorum.

Berry, Donald A. (1996). *Statistics: A Bayesian Perspective*. Belmont CA: Wadsworth.

Blackwell, David (1969). *Basic Statistics*. New York: McGraw-Hill.

Box, George E. P., and George C. Tiao (1973). *Bayesian Inference in Statistical Analysis*. Reading, MA: Addison-Wesley.

Broemeling, Lyle D. (1985). *Bayesian Analysis of Linear Models*. New York: Marcel Dekker.

Buck, Caitlin E., William G. Cavanagh, and Clifford D. Litton (1996). *Bayesian Approach to Interpreting Archaeological Data*. New York: Wiley.

Carlin, Bradley, and Thomas A. Louis (1998). *Bayes and Empirical Bayes Methods for Data Analysis*. Boca Raton, FL: Chapman & Hall/CRC Press.

Dale, Andrew I. (1991). *A History of Inverse Probability from Thomas Bayes to Karl Pearson*. New York: Springer-Verlag.

De Finetti, Bruno (1973). *Theory of Probability*, Vols. 1 and 2, translated from the Italian by Antonio Machi and Adrian Smith. New York: Wiley.

DeGroot, Morris H. (1970). *Optimal Statistical Decisions*. New York: McGraw-Hill.

——— (1975). *Probability and Statistics*. Reading, MA: Addison-Wesley.

Dey, Dipak, Peter Muller, and Sinha Debajyoti (Eds.) (1998). *Practical Nonparametric and Semiparametric Bayesian Statistics*, Vol. 133 of Lecture Notes in Statistics. New York: Springer-Verlag.

Fisher, R. A. (1922). "On the Mathematical Foundations of Theoretical Statistics," *Philosophical Transactions the Royal Society of London*, **A222**:309–368.

——— (1935). "Statistical Tests," *Nature*, **136**:474.

——— (1936). "Has Mendel's Work Been Rediscovered?" *Annals of Science*, **1**:115–137.

——— (1970). *Statistical Methods for Research Workers*, 14th ed. Edinburgh: Oliver & Boyd (first published in 1925).

——— (1947). *The Design of Experiments*, 4th ed. Edinburgh: Oliver & Boyd (first published in 1926).

——— (1956). *Statistical Methods and Statistical Inference*. Edinburgh: Oliver & Boyd.

Florens, Jean-Pierre, Michel Mouchart, and Jean-Marie Roin (1990). *Elements of Bayesian Statistics*. New York: Marcel Dekker.

Gamerman, Dani (1997). *Markov Chain Monte Carlo: Stochastic Simulation for Bayesian Inference*. Boca Raton, FL: CRC Press.

Gatsonis, C., J. S. Hodges, R. E. Kass, and N. D. Singpurwalla (Eds.) (1993). *Case Studies in Bayesian Statistics*. New York: Springer-Verlag.

——(1995). *Case Studies in Bayesian Statistics*, Vol. II. New York: Springer-Verlag.

Gatsonis, C., J. S. Hodges, R. E. Kass, R. McCulloch, P. Rossi, and N. D. Singpurwalla (Eds.) (1997). *Case Studies in Bayesian Statistics*, Vol. III. New York: Springer-Verlag.

Gatsonis, C., B. Carlin, A. Gelman, M. West, R. E. Kass, A. Carriquiry, and I. Verdinelli (Eds.) (1999). *Case Studies in Bayesian Statistics*, Vol. IV. New York: Springer-Verlag.

Geisser, Seymour (1993). *Predictive Inference: An Introduction*. New York: Chapman & Hall.

Gelman, Andrew, John B. Carlin, Hal S. Stern, and Donald B. Rubin (1995). *Bayesian Data Analysis*. London: Chapman & Hall.

Gilks, W. R., S. Richardson, and D. J. Spiegelhalter (Eds.) (1996). *Markov Chain Monte Carlo in Practice*. Boca Raton, FL: CRC Press.

Gillispie, Charles Coulston (1960). *The Edge of Objectivity: An Essay in the History of Scientific Ideas*. Princeton, NJ: Princeton University Press.

Good, I. J. (1950). *Probability and the Weighting of Evidence*. London: Charles Griffin.

——(1965). *The Estimation of Probabilities: An Essay on Modern Bayesian Methods*, Research Monograph 30. Cambridge, MA: MIT Press.

——(1974). *Information, Weight of Evidence, the Singularity Between Probability Measures and Signal Detection*. New York: Springer-Verlag.

——(1983). *Good Thinking: The Foundations of Probability and Its Applications*. Minneapolis, MN: University of Minnesota Press.

Hadamard, Jacques (1945). *An Essay on the Psychology of Invention in the Mathematical Field*. Princeton, NJ: Princeton University Press.

Hartigan, J. A. (1983). *Bayes Theory*. New York: Springer-Verlag.

Howson, Colin, and Peter Urbach (1989). *Scientific Reasoning: The Bayesian Approach*. La Salle, IL: Open Court.

Jaynes, E. T. (1983). In R. D. Rosenkrantz (Ed.), *Papers on Probability, Statistics, and Statistical Physics of E. T. Jaynes*. Dordrecht, The Netherlands: D. Reidel.

Jeffreys, Harold (1939, 1948, 1961). *Theory of Probability*, 1st 2nd, and 3rd eds. Oxford: Oxford University Press.

Kadane, Joseph B., Mark J. Schervish, and Teddy Seidenfeld (1999). *Rethinking the Foundations of Statistics*. Cambridge: Cambridge University Press.

Keynes, John Maynard (1921). *A Treatise on Probability*. London: Macmillan.

Kuhn, Thomas S. (1962). *The Structure of Scientific Revolutions*, 2nd ed., enlarged. Chicago: University of Chicago Press.

Lad, Frank (1996). *Operational Subjective Statistical Methods: A Mathematical, Philosophical, and Historical Introduction*. New York: Wiley-Interscience.

Lakatos, I. (1968). "Criticism and Methodology of Scientific Research Programmes," *Proceedings of the Aristotelian Society*, **69**:149–186.

——(1970). "Falsification and the Methodology of Scientific Research Programmes," in I. Lakatos and A. Musgrave (Eds.), *Criticism and the Growth of Knowledge*. Cambridge: Cambridge University Press.

————(1974). "Popper on Demarcation and Induction," Chapter 5 in P. A. Schilpp (Ed.), *The Philosophy of Karl Popper*, Vol. 2. LaSalle, IL: Open Court.

————(1978). In J. Worrall, and G. Currie (Eds.), *Philosophical Papers*, 2 vols. Cambridge: Cambridge University Press.

Laplace, P. S. (1774). "Mémoire Sur la Probabilité des Causes par les Evénements," *Memories de l'Academic Royale Sciences Presents Par Divers Savans*, **6**:621–656.

Lee, Peter M. (1997). *Bayesian Statistics: An Introduction*, 2nd ed. New York: Wiley.

Leonard, Thomas, and John S. J. Hsu (1999). *Bayesian Methods: An Analysis for Statisticians and Interdisciplinary Researchers*. Cambridge: Cambridge University Press.

Lindley, Dennis V. (1965). *Introduction to Probability and Statistics*, 2 vols. (Part 1: Probability; Part 2: Inference). Cambridge: Cambridge University Press.

————(1972). *Bayesian Statistics: A Review*. Philadelphia: Society for Industrial and Applied Mathematics.

Malakoff, David (1999). "Bayes Offers a 'New' Way to Make Sense of Numbers," *Science*, **286**:1460–1464.

Mathews, Jon, and R. L. Walker (1965). *Mathematical Methods of Physics*. New York: W. A. Benjamin, Inc.

Matthews, Robert A. J. (1998a). *Fact Versus Factions: The Use and Abuse of Subjectivity in Scientific Research*, ESEF Working Paper 2/98, European Science and Environment Forum, Cambridge, September.

————(1998b). "Flukes and Flaws," *Prospect*, Issue 35, pp. 20–24.

Meeden, G., and Malay Ghosh (1997). *Bayesian Methods for Finite Population Sampling*. Boca Raton, FL: CRC Press.

Mockus, Jonas, William Eddy, Audris Mockus, Linas Mockus, and G. V. Reklaitis (1997). *Bayesian Heuristic Approach to Discrete and Global Optimization: Algorithms, Visualization, Software, and Applications*. Boston: Kluwer Academic Publishers.

Mosteller, Frederick, and David L. Wallace (1984). *Applied Bayesian and Classical Inference*, 2nd ed. New York: Springer-Verlag.

Muller, Peter, and Brani Vidakovic (Eds.) (1999). *Bayesian Inference in Wavelet-Based Models*, Vol. 141 of Lecture Notes in Statistics. New York: Springer-Verlag.

Neal, Radford M. (1996). *Bayesian Learning for Neural Networks*, Vol. 118 of Lecture Notes in Statistics. New York: Springer-Verlag.

Neyman, Jerzy, and Lucian M. Le Cam (Eds.) (1965). *Bernoulli (1713), Bayes (1763), Laplace (1813)*, Anniversary Volume. New York: Springer-Verlag.

Neyman, J., and E. S. Pearson (1928). "On the Use and the Interpretation of Certain Test Criteria for Purposes of Statistical Inference," *Biometrika*, **20**:175–240, Part I; 263–294, Part II.

————(1933). "On the Problem of the Most Efficient Tests of Statistical Hypotheses," *Philosophical Transactions of the Royal Society*, **231A**:289–337.

O'Hagen, Anthony (1994). "Bayesian Inference," in *Kendall's Advanced Theory of Statistics*, Vol. 2B. London: Edward Arnold/Hodder Headline.

Polanyi, Michael (1958, 1962). *Personal Knowledge: Towards a Post-Critical Philosophy*. Chicago: University of Chicago Press (corrected edition, 1962).

Popper, Sir Karl (1956, 1982). *Quantum Theory and the Schism in Physics*, from the Postscript to *The Logic of Scientific Discovery*, W. W. Bartley III (Ed.). Totowa, NJ: Rowman & Littlefield.

———(1959a). "The Propensity Interpretation of Probability," *British Journal for the Philosophy of Science*, **10**:25–42.

———(1959b). *The Logic of Scientific Discovery*. London: Hutchinson; New York: Harper & Row.

———(1963). *Conjectures and Refutations*. New York: Routledge & Kegan.

Press, S. James (1989). *Bayesian Statistics: Principles, Models and Applications*. New York: Wiley.

Raiffa, Howard, and Robert Schlaifer (1961). *Applied Statistical Decision Theory*. Boston: Harvard Graduate School of Business Administration.

Ramsey, Frank Plimpton (1926). "Truth and Probability," in R. B. Braithwaite (Ed.), *The Foundations of Mathematics and Other Logical Essays*. New York: Humanities Press, 1950. Reprinted in Henry E. Kyburg and Howard F. Smokler (Eds.), *Studies in Subjective Probability*, 1980. Huntington, NY: Robert E. Krieger, pp. 23–52.

Roberts, Royston M. (1989). *Serendipity: Accidental Discoveries in Science*. New York: Wiley.

Savage, L. J. (1954). *The Foundations of Statistics*. New York: Wiley.

———(1981). *The Writings of Leonard Jimmie Savage—A Memorial Selection*. Washington, DC: American Statistical Association and Institute of Mathematical Statistics.

Schlaifer, Robert (1961). *Introduction to Statistics for Business Decisions*. New York: McGraw-Hill.

Schmitt, Samuel A. (1969). *Measuring Uncertainty: An Elementary Introduction to Bayesian Statistics*. Reading, MA: Addison-Wesley.

Stigler, S. M. (1986). *The History of Statistics*. Cambridge, MA: Harvard University Press.

Tanner, Martin A. (1996). *Tools for Statistical Inference: Methods for the Exploration of Posterior Distributions and Likelihood Functions*, 3rd ed. New York: Springer-Verlag.

Viertl, R. (Ed.) (1987). *Probability and Bayesian Statistics*. New York: Plenum Press.

Watson, James D., and Francis H. C. Crick (1953). "A Structure for Deoxyribose Nucleic Acid." *Nature*, **171**:737.

West, Mike, and Jeff Harrison (1997). *Bayesian Forecasting and Dynamic Models*, 2nd ed. New York: Springer-Verlag.

Winkler, Robert L. (1972). *An Introduction to Bayesian Inference and Decision*. New York: Holt, Rinehart and Winston.

Wolpert, Lewis (1992). *The Unnatural Nature of Science*. Boston: Harvard University Press.

Zellner, Arnold (1971). *An Introduction to Bayesian Inference in Econometrics*. New York: Wiley.

———(1997). *Bayesian Analysis in Econometrics and Statistics: The Zellner View and Papers*. Cheltenham, UK: Edward Elgar.

References by Field of Application for Bayesian Statistical Science

A considerable amount of Bayesian statistical inference procedures that formally admit meaningful prior information in the scientific process of data analysis have had to await the advent of modern computer methods of analysis. This did not really happen in any major way until about the last couple of decades of the twentieth century. Moreover, at approximately the same time, numerical methods for implementing Bayesian solutions to complex applied problems were greatly augmented by the introduction of Markov Chain Monte Carlo (MCMC) methods (Gelfand and Smith, 1990; Geman and Geman, 1984). Since the arrival of these developments, Bayesian methods have been very usefully applied to problems in many fields, and the general approach has generated much new research.

A.1 BAYESIAN METHODOLOGY

Some books and articles summarizing Bayesian methodology in various fields are given below.

1. *Anthropology and archaeology:* Buck et al. (1996).
2. *Econometrics:* Zellner (1971).
3. *Economics:* Dorfman (1997).
4. *Evaluation research:* Pollard (1986).
5. *Mutual fund management:* Hulbert (1999).
6. *Physics and engineering:* There is a series of books on maximum entropy and Bayesian methods published by Kluwer; a few are given in the References at the end of this appendix, listed under "Kluwer." See also

Stone et al. (1999) for target tracking, O'Ruanaidh and Fitzgerald (1996) for signal processing, and Jaynes (1983) for physics.

7. *Reliability:* Sander and Badoux (1991).

A.2 APPLICATIONS

For the convenience of readers in various fields, we have organized and presented immediately below, by discipline, citations of some technical papers that use the Bayesian approach to find solutions to real problems. Sometimes a paper really belongs to several of our discipline categories but is only listed in one. These citations have been drawn from the following four volumes, entitled *Case Studies in Bayesian Statistics*, published by Springer-Verlag (New York). The first volume (which we shall refer to as volume I, although it actually has no volume number) published in 1993, and Volume II, published in 1995, were edited by Gatsonis, Hodges, Kass, and Singpurwalla; Volume III, published in 1997, was edited by Gatsonis, Hodges, Kass, McCulloch, Rossi, and Singpurwalla; Volume IV, published in 1999, was edited by Gatsonis, Carlin, Gelman, West, Kass, Carriquiry, and Verdinelli.

Biochemistry

Etzioni, Ruth, and Bradley P. Carlin, "Bayesian Analysis of the Ames Salmonella/Microsome Assay," Vol. I, p. 311.

Business Economics

Nandram, B., and J. Sedransk, "Bayesian Inference for the Mean of a Stratified Population When There Are Order Restrictions," Vol. II, p. 309.

Ecology

Lad, Frank, and Mark W. Brabyn, "Synchronicity of Whale Strandings with Phases of the Moon," Vol. I, p. 362.

Raftery, Adrian E., and Judith E. Zeh, "Estimation of Bowhead Whale, *Balaena mysticetus*, Population Size," Vol. I, p. 163.

Wolfson, L. J., J. B. Kadane, and M. J. Small, "A Subjective Bayesian Approach to Environmental Sampling," Vol. III, p. 457.

Wolpert, Robert L., Laura J. Steinberg, and Kenneth H. Reckhow, "Bayesian Decision Support Using Environmental Transport-and-Fate Models," Vol. I, p. 241.

Engineering

Andrews, Richard W., James O. Berger, and Murray H. Smith, "Bayesian Estimation of Fuel Economy Potential Due to Technology Improvements," Vol. I, p. 1.

Jewell, William S., and Shrane-Koung Chou, "Predicting Coproduct Yields in Microchip Fabrication," Vol. I, p. 351.

O'Hagan, Anthony, and Frank S. Wells, "Use of Prior Information to Estimate Costs in a Sewerage Operation," Vol. I, p. 118.

Clyde, Merlise, Peter Muller, and Giovanni Parmigiani, "Optimal Design for Heart Defibrillators," Vol. II, p. 278.

Craig, Bruce A., and Michael A. Newton, "Modeling the History of Diabetic Retinopathy," Vol. III, p. 305.

Crawford, Sybil L., William G. Johnson, and Nan M. Laird, "Bayes Analysis of Model-Based Methods for Non-ignorable Nonresponse in the Harvard Medical Practice Survey," Vol. I, p. 78.

Crawford, Sybil L., Sharon L. Tennstedt, and John B. McKinlay, "Longitudinal Care Patterns for Disabled Elders: A Bayesian Analysis of Missing Data," Vol. II, p. 293.

Flournoy, Nancy, "A Clinical Experiment in Bone Marrow Transplantation: Estimating a Percentage Point of a Quantal Response Curve," Vol. I, p. 324.

Normand, Sharon-Lise T., Mark E. Glickman, and Thomas J. Ryan, "Modeling Mortality Rates for Elderly Heart Attack Patients: Profiling Hospitals in the Cooperative Cardiovascular Project," Vol. III, p. 155.

Parmigiani, Giovanni, and Mark S. Kamlet, "A Cost-Utility Analysis of Alternative Strategies in Screening for Breast Cancer," Vol. I, p. 390.

Meteorology

Smith, R. L., and P. J. Robinson, "A Bayesian Approach to the Modeling of Spatial-Temporal Precipitation Data," Vol. III, p. 237.

Neurophysiology

Genovese, C. R., and J. A. Sweeney: "Functional Connectivity in the Cortical Circuits Subserving Eye Movements," Vol. IV, p. 59.

West, Mike, and Guoliang Cao, "Assessing Mechanisms of Neural Synaptic Activity," Vol. I, p. 416.

Oncology/Genetics

Adak, S., and A. Sarkar, "Longitudinal Modeling of the Side Effects of Radiation Therapy," Vol. IV, p. 269.

Iversen, E. S. Jr., G. Parmagiani, and D. A. Berry, "Validating Bayesian Prediction Models: A Case Study in Genetic Susceptibility to Breast Cancer," Vol. IV, p. 321.

Palmer, J. L., and P. Muller: "Population Models for Hematologic Models," Vol. IV, p. 355.

Parmigiani, G., D. A. Berry, E. S. Iversen, Jr., P. Muller, J. M. Schildkraut, and E. P. Winer, "Modeling Risk of Breast Cancer and Decisions About Genetic Testing," Vol. IV, p. 133.

Slate, E. H., and L. C. Clark: "Using PSA to Detect Prostate Cancer Onset: An Application of Bayesian Retrospective and Prospective Changepoint Identification," Vol. IV, p. 395.

Slate, Elizabeth H., and Kathleen A. Crohin, "Changepoint Modeling of Longitudinal PSA as a Biomarker for Prostate Cancer," Vol. III, p. 435.

Pharmacology/Pharmacokinetics

Paddock, S., M. West, S. S. Young, and M. Clyde, "Mixture Models in the Exploration of Structure–Activity Relationships in Drug Design," Vol. IV, p. 339.

Wakefield, J., Aarons, L., and A. Racine-Poon: "The Bayesian Approach to Population Pharmacokinetic/Pharmacodynamic Modeling," Vol. IV, p. 205.

Physiology
Brown, Emery N., and Adam Sapirstein, "A Bayesian Model for Organ Blood Flow Measurement with Colored Microspheres," Vol. II, p. 1.

Political Science
Bernardo, José M., "Probing Public Opinion: The State of Valencia," Vol. III, p. 3.

Psychiatry
Erkanli, Alaattin, Refik Soyer, and Dalene Stangl, "Hierarchical Bayesian Analysis for Prevalence Estimation," Vol. III, p. 325.

Public Health
Aguilar, O., and M. West: "Analysis of Hospital Quality Monitors Using Hierarchical Time Series Models," Vol. IV, p. 287.

Belin, Thomas R., Robert M. Elashoff, Kwan-Moon Leung, Rosane Nisenbaum, Roshan Bastani, Kiumarss Nasseri, and Annette Maxwell, "Combining Information from Multiple Sources in the Analysis of a Non-equivalent Control Group Design," Vol. II, p. 241.

Blattenberger, Gail, and Richard Fowles, "Road Closure: Combining Data and Expert Opinion," Vol. II, p. 261.

Carlin, Bradley P., Kathryn M. Chaloner, Thomas A. Louis, and Frank S. Rhame, "Elicitation, Monitoring, and Analysis for an AIDS Clinical Trial," Vol. II, p. 48.

Malec, Donald, Joseph Sedransk, and Linda Tompkins, "Bayesian Predictive Inference for Small Areas for Binary Variables in the National Health Interview Survey," Vol. I, p. 377.

Radiology
Johnson, Valen, James Bowsher, Ronald Jaszczak, and Timothy Turkington, "Analysis and Reconstruction of Medical Images Using Prior Information," Vol. II, p. 149.

REFERENCES

Buck, Caitlin E., William G. Cavanagh, and Clifford D. Litton (1996). *Bayesian Approach to Interpreting Archaeological Data.* New York: John Wiley & Sons.

Dorfman, Jeffrey H. (1997). *Bayesian Economics Through Numerical Methods: A Guide to Econometrics and Decision-Making with Prior Information.* New York: Springer-Verlag.

Gelfand, A. E., and A. F. M. Smith (1990). "Sampling Based Approaches to Calculating Marginal Densities," *Journal of the American Statistical Association*, **85**:398–409.

Geman, S., and D. Geman (1986). "Stochastic Relaxation, Gibbs Distributions and the Bayesian Restoration of Images," *IEEE Transactions on Pattern Analysis and Machine Intelligence*, **6**:721–741.

Hulbert, Mark (1999). "Are Fund Managers Irrelevant? An 18th Century Theory Suggests Not," Business Section, *New York Times*, October 10. p. 26.

Jaynes, E. T. (1983). In R. D. Rosenkrantz (Ed.), *Papers on Probability, Statistics, and Statistical Physics of E. T. Jaynes*. Dordrecht, The Netherlands: D. Reidel.

Kluwer Academic Publishers (1982). *Maximum-Entropy and Bayesian Spectral Analysis and Estimation Problems*, C. Ray Smith, and Gary J. Erickson (Eds.).

Kluwer Academic Publishers (1991). *Maximum Entropy and Bayesian Methods, Proceedings of the 11th International Workshop on Maximum Entropy and Bayesian Methods*, Seattle, WA.

Kluwer Academic Publishers (1992). *Proceedings of the 12th International Workshop on Maximum Entropy and Bayesian Methods*, Paris, A. Mohammad-Djafari and G. Demoment (Eds.).

Kluwer Academic Publishers (1994). *Proceedings of the 14th International Workshop on Maximum Entropy and Bayesian Methods*, Cambridge, John Skilling and Sibusiso Sibisi (Eds.).

Kluwer Academic Publishers (1995). *Proceedings of the 15th International Workshop on Maximum Entropy and Bayesian Methods, Santa Fe*, NM, Kenneth M. Hanson and Richard N. Silver (Eds.).

Kluwer Academic Publishers (1997). *Proceedings of the 17th International Workshop on Maximum Entropy and Bayesian Methods*, Boise, ID, Gary J. Erickson, Joshua T. Rychert, and C. Ray Smith (Eds.)

Kluwer Academic Publishers (1999). *Maximum Entropy and Bayesian Methods, Proceedings of the 18th International Workshop on Maximum Entropy and Bayesian Methods*.

Lad, Frank (1996). *Operational Subjective Statistical Methods: A Mathematical, Philosophical, and Historical Introduction*. New York: Wiley-Interscience.

O'Ruanaidh, Joseph J. K., and William J. Fitzgerald (1996). *Numerical Bayesian Methods Applied to Signal Processing*. New York: Springer-Verlag.

Pollard, William E. (1986). *Bayesian Statistics for Evaluation Research: An Introduction*. Beverly Hills, CA: Sage Publications.

Sander, P., and R. Badoux (Eds.) (1991). *Bayesian Methods in Reliability*. Boston: Kluwer Academic Publishers.

Smith, C. Ray (Ed.) (1985). *Maximum-Entropy and Bayesian Methods in Inverse Problems*, Vol. 14 of Fundamental Theories of Physics. Dordrecht, The Netherlands: D. Reidel.

Stone, Lawrence D., Carl A. Barlow, and Thomas L. Corwin (1999). *Bayesian Multiple Target Tracking*. Norwood, MA: Artech House.

Zellner, Arnold (1971). *An Introduction to Bayesian Inference in Econometrics*. New York: John Wiley & Sons.

Bibliography

Abraham, Hilda C., and Ernst L. Freud (Eds.) (1965). *A Psycho-Analytic Dialogue: The Letters of Sigmund Freud and Karl Abraham, 1907–1926*. London: Hogarth Press.

Agassi, Joseph (1971). *Faraday as Natural Philosopher*. Chicago: University of Chicago Press.

Albert, James H. (1996). *Bayesian Computation Using Minitab*. Belmont, CA: Wadsworth.

Alexander, F., and Sheldon Selesnick (1966). *The History of Psychiatry*. London: Hogarth Press.

Altman, Lawrence K. (1995). "Revisionist History Sees Pasteur as Liar Who Stole Rival's Ideas," *New York Times*, May 16, pp. C1, C3.

Andreas-Salome, Lou (1965). *The Freud Journal of Lou Andreas-Salome* (English translation of *In der Schule bei Freud: Tage-buch eines Jahres, 1912–1913*). London: Hogarth Press.

Antleman, Gordon (1997). *Elementary Bayesian Statistics*. Albert Madansky and Robert E. McCulloch (Eds.). Cheltenham, UK: Edward Elgar.

Aristotle (1943). Introduction by Louise Ropes Loomis. New York: Gramercy Books.

Asimov, Isaac (1989). *Asimov's Chronology of Science and Discovery*. New York: Harper & Row.

Bakan, David (1958a). "Moses in the Thought of Freud," *Commentary*, October, pp. 322–331.

———(1958b). *Sigmund Freud and the Jewish Mystical Tradition*. Princeton, NJ: D. Van Nostrand.

Barnes, Johnathan (1982). *Aristotle*. Oxford: Oxford University Press.

———(Ed.) (1995). *The Cambridge Companion to Aristotle*. Cambridge: Cambridge University Press.

Barnett, Lincoln (1948). *The Universe and Dr. Einstein*. New York: New American Library of World Literature, Mentor Books.

Basu, D. (1988). *Statistical Information and Likelihood: A Collection of Critical Essays*, Vol. 45 of Lecture Notes in Statistics. New York: Springer-Verlag.

Bauer, Henry H. (1992). *Scientific Literacy and the Myth of the Scientific Method*, Urbana, IL: University of Illinois Press.

Bayes, Thomas (1763). "An Essay Towards Solving a Problem in the Doctrine of Chances," *Philosophical Transactions of the Royal Society of London*, **53**:370–418.

Beadle, G. (1967). "Mendelism," pp. 335–350 in A. Brink (Ed.), *Heritage from Mendel*. Madison, WI: University of Wisconsin Press.

Bell, E. T. (1937). *Men of Mathematics*. New York: Simon & Schuster.

Bell, Robert Ivan (1992). *Impure Science: Fraud, Compromise, and Political Influence in Scientific Research*, New York: John Wiley & Sons.

Ben-Yehuda, Nachum (1985). *Deviance and Moral Boundaries*. Chicago: University of Chicago Press.

———(1986). "Deviance in Science: Towards the Criminology of Science," *British Journal of Criminology*, **26**:1–27.

Berger, James O. (1985). *Statistical Decision Theory and Bayesian Analysis*. New York: Springer-Verlag.

Berger, James O., and Donald A. Berry (1988). "Statistical Analysis and the Illusion of Objectivity," *American Scientist*, **76**:159–165, March–April.

Berman, Morris (1978). *Social Change and Scientific Organization: The Royal Institution, 1799–1844*. Ithaca, NY: Cornell University Press.

Bernardo, José M., and Adrian F. M. Smith (1994). *Bayesian Theory*, New York: John Wiley & Sons.

Bernoulli, James (1713). *Ars Conjectandi* (The Art of Conjecturing), Book 4. Basileae: Impensis Thurnisiorum.

Berry, Donald A. (1996). *Statistics: A Bayesian Perspective*. Belmont, CA: Wadsworth.

Biagioli, Mario (1993). *Galileo: Courtier*. Chicago: University of Chicago Press.

Blackwell, David (1969). *Basic Statistics*. New York: Mc Graw Hill Book Co.

Bolton, Sarah Knowles (1960). *Famous Men of Science*, revised by Barbara Lovett Cline. New York: Thomas Y. Crowell.

Bonaparte, Marie, Anna Freud, and Ernst Kris (Eds.) (1887–1902). *Sigmund Freud, the Origins of Psychoanalysis: Letters to Wilhelm Fliess, Drafts and Notes*, English translation by Eric Mosbacher and James Strachey. New York: Basic Books.

Born, Max (1916). *Physica Acta*. Hebrew University, Israel: The Einstein Archives, p. 253.

Botting, Douglas (1973). *Humboldt and the Cosmos*. New York: Harper & Row.

Bottome, Phyllis, and Alfred Adler (1946). *Apostle of Freedom*, 2nd ed. London: Faber & Faber.

Bourdieu, Pierre (1991). "The Peculiar History of Scientific Reason," *Sociological Forum*, **6**:3–26.

Bower, Bruce (1998). "Objective Visions: Historians Track the Rise and Times of Scientific Objectivity," *Science News*, **154**:360–362, Dec. 5.

Box, George E. P., and George C. Tiao (1973). *Bayesian Inference in Statistical Analysis*. Reading, MA: Addison-Wesley.

Broad, William J. (1990). "After 400 Years, a Challenge to Kepler: He Fabricated His Data, Scholar Says," Science Times, *New York Times*, January 23, pp. B5, B7.

Broad, William, and Nicholas Wade (1982). *Betrayers of the Truth*, New York: Simon & Schuster.

Broemeling, Lyle D. (1985). *Bayesian Analysis of Linear Models.* New York: Marcel Dekker.

Brome, Vincent (1967). *Freud and His Early Circle*, New York: Morrow.

Brooke, John Hedley (1991). *Science and Religion: Some Historical Perspectives.* Cambridge: Cambridge University Press.

Bruhns, Karl (Ed.) (1873). *Life of Alexander von Humboldt Compiled in Commemoration of the Centenary of His Birth* by J. Lowenberg, Robert Avé-Lallemant, and Alfred Dove, 2 vols., translated by Jane and Caroline Lassell. London: Longmans, Green.

Buck, Caitlin E., William G. Cavanagh, and Clifford D. Litton (1996). *Bayesian Approach to Interpreting Archaeological Data*, New York: John Wiley & Sons.

Burt, C. L. (1909). "Experimental Tests of General Intelligence," *British Journal of Psychology*, **3**:94–177.

———(1961). "Intelligence and Social Mobility," *British Journal of Statistical Psychology*, **14**:3–24.

Calaprice, Alice (1996). *The Quotable Einstein.* Princeton, NJ: Princeton University Press.

Cantor, Geoffrey N. (1985). "Reading the Book of Nature: The Relation Between Faraday's Religion and His Science," pp. 69–81 in David Gooding and Frank A. J. L. James (Eds.), *Faraday Rediscovered: Essays on the Life and Work of Michael Faraday, 1791–1867.* Bassingstoke, Hants, UK: Macmillan. (Published in the United States and Canada by Stockton Press, New York.)

Carlin, Bradley, and Thomas A. Louis (1998). *Bayes and Empirical Bayes Methods for Data Analysis.* Boca Raton, FL: Chapman & Hall/CRC Press.

Casti, John (1996). "Confronting Science's Logical Limits," *Scientific American*, October, pp. 102–105.

Caton, Hiram (Ed.) (1990). *The Samoa Reader: Anthropologists Take Stock.* Lanham, MD: University Press of America.

Chalmers, Alan (1990). *Science and Its Fabrication.* Minneapolis, MN: University of Minnesota Press.

Chambers Concise Dictionary of Scientists (1989). Cambridge: W & R Chambers and the Press Syndicate of the University of Cambridge.

Christianson, Gale E. (1984). *In the Presence of the Creator: Isaac Newton and His Times.* New York: Free Press.

Clark, Ronald W. (1971). *Einstein: The Life and Times.* New York: Avon Books.

Clegg, Michael (1997). Professor of Genetics, Department of Botany and Plant Sciences, University of California, Riverside, personal communication, March.

Cline, Barbara Lovett (1960). See Bolton (1960).

Collins, Harry, and Trevor Pinch (1993). *The Golem: What Everyone Should Know About Science.* Cambridge: Cambridge University Press.

Conant, James Bryant (1956). *The Overthrow of the Phlogiston Theory*, Vol. 2 of Harvard Case Studies in Experimental Science. Cambridge, MA: Harvard University Press.

Correns, Carl G. (1900). *Mendel's Regel über das Verhalten der Nachkommenschaft der Rassenbastarde," Botanischen Gesellschaft Berichte*, **18**:158–168.

Costigan, Giovanni (1968). *Sigmund Freud: A Short Biography*. New York: Collier Books.

Crowther, J. G. (1936). *Men of Science*. New York: W.W. Norton.

———(1995). *Six Great Scientists: Copernicus, Galileo, Newton, Darwin, Marie Curie, Einstein*. New York: Barnes & Noble.

Curie, Eve (1939). *Madame Curie: A Biography by Eve Curie*, translated by Vincent Sheean. New York: Doubleday, Doran.

Dagognet, Francois (1967). *Méthodes et Doctrines dans l'Oeuvre de Pasteur*. Paris: Presses Universitaires.

Dale, Andrew I. (1991). *A History of Inverse Probability from Thomas Bayes to Karl Pearson*. New York: Springer-Verlag.

Darbishire, A. D. (1911). *Breeding and the Mendelian Discovery*. London: Cassell and Company.

Darwin, Francis (Ed.) (1887). *Life and Letters of Charles Darwin*, 3 vol. New York: Appleton.

———(1903). *More Letters of Charles Darwin*, 2 vols. London: John Murray, Vol. 1.

Darwin, Francis, and A. C. Seward (Eds.) (1903). *More Letters: Letters of Charles Darwin*, 2 vols. London: John Murray.

Daston, Lorraine J., and Peter Galison (1992). "The Image of Objectivity," *Representations*, **40**:81–128.

Davis, Kenneth S. (1966). *The Cautionary Scientists: Priestley, Lavoisier, and the Founding of Modern Chemistry*. New York: G.P. Putnam's Sons.

Dawes, Robyn (1994). *House of Cards: Psychology and Psychotherapy Built on Myth*. New York: Free Press.

De Beer, G. (1964). "Mendel, Darwin, and Fisher," *Notes and Records of the Royal Society, London*, **19**(2):192–226.

De Beer, Sir Gavin (Ed.) (1974). *Darwin and Huxley: Autobiographies*. London: Oxford University Press.

De Finetti, Bruno (1973). *Theory of Probability*, Vols. 1 and 2, translated from the Italian by Antonio Machi and Adrian Smith. New York: John Wiley & Sons.

DeGroot, Morris H. (1970). *Optimal Statistical Decisions*. New York: McGraw-Hill.

———(1975). *Probability and Statistics*. Reading, MA: Addison-Wesley.

DeKruif, Paul (1953). *Microbe Hunters*. New York: Harbrace Paperback Library.

de Terra, Helmut (1955). *Humboldt: The Life and Times of Alexander von Humboldt, 1769–1859*. New York: Alfred A. Knopf.

Devlin, Bernie, Daniel P. Resnick, Stephen E. Fienberg, and Kathryn Roeder (Eds.) (1997). *Intelligence, Genes and Success: Scientists Respond to The Bell Curve*. New York: Springer-Verlag.

De Vries, Hugo (1900). "Sur la Loi de Disjonction des Hybrides," *Comptes Rendus de l'Academie des Sciences, Paris*, **130**:845–847.

Dey, Dipak, Peter Muller, and Sinha Debajyoti (Eds.) (1998). *Practical Nonparametric and Semiparametric Bayesian Statistics*, Vol. 133 of Lecture Notes in Statistics. New York: Springer-Verlag.

Dolin, Arnold (1960). *Great Men of Science.* New York: Hart.

Donahue, W. H. (1988). "Kepler's Fabricated Figures: Covering Up the Mess in *The New Astronomy*," *Journal for the History of Astronomy*, **19**:216–237.

——(Trans.) (1992). *The New Astronomy* by Johannes Kepler. Cambridge: Cambridge University Press.

Donovan, Arthur (1993). *Antoine Lavoisier.* Cambridge, MA: Blackwell.

Dorfman, D. D. (1978). "The Cyril Burt Question: New Findings," *Science*, **201**:1177–1186.

——(1979). "Burt's Tables," Letters to the Editor, *Science*, **204**:246–254.

Dorfman, Jeffrey H. (1997). *Bayesian Economics Through Numerical Methods: A Guide to Econometrics and Decision-Making with Prior Information.* New York: Springer-Verlag.

Drake, Stillman (1990). *Galileo: Pioneer Scientist.* Toronto: University of Toronto Press.

Dreger, R. M. (Ed.) (1972). *Multivariate Personality Research.* Baton Rouge, LA: Claitor's Publishing Division.

Dubos, René (1950, 1960, 1976). *Louis Pasteur: Free Lance of Science.* New York: Scribner's, Da Capo Press.

Duclaux, Emile (1895). "Le Laboratoire de Monsieur Pasteur a l'Ecole Normale," *Revue Scientifique*, 4th ser., **15**:449–454; reprinted in *Le Centenaire de l'Ecole Normale, 1795–1895.* Paris: Ecole Normale.

Duclaux, Emile (1896). *Pasteur: Histoire d'un Esprit.* Sceaux, France: Charaire. English translation by E. F. Smith and F. Hedges as *Pasteur: The History of a Mind* (1920). Philadelphia: W.B. Saunders.

Dunn, L. C. (1965). "Mendel: His Work and His Place in History," *Proceedings of the American Philosophical Society*, **109**:189–198.

Edwards, A. W. F. (1986). "Are Mendel's Results Really Too Close?" *Biological Review*, **61**:295–312.

Einstein, Albert (1967). "Foreword" to Galileo Galilei *Dialogue Concerning the Two Chief World Systems—Ptolemaic and Copernican*, 2nd ed., translated by Stillman Drake. Berkeley, CA: University of California Press. The foreword is dated July 1952 and is an authorized translation from the German by Sonja Bargmann.

——(1993). *Einstein on Humanism.* Secaucus, NJ: Carol Publishing Group, Citadel Press.

Encyclopaedia Britannica (1975).

——(1996). "Aristotle and Aristotelianism: Assessment and Nature of Aristotelianism," compact disk version.

Encyclopaedia Britannica Micropedia (1975).

Encyclopedia Americana (1958). Vol. 20.

Eysenck, Hans, and Glenn D. Wilson (1973). *The Experimental Study of Freudian Theories.* London: Methuen.

Faraday, Michael (1816a). *Chemistry Lectures.* London: Institute of Engineers.

——(1816b). "Observations on the Inertia of the Mind," in *Commonplace Book, 1816–1846.* London: Institute of Engineers.

——(1839–1855). *Experimental Researches in Electricity.* Reprinted from the *Philosophical Transactions* of 1831–1852 with other electrical papers from the *Quarterly*

Journal of Science, Philosophical Magazine, Proceedings of the Royal Institution, 3 vols. London.

——— (1854). "Observations on Mental Education," pp. 39–88 in *Lectures on Education Delivered at the Royal Institution of Great Britain.* London: J. W. Parker and Son. Reprinted in *Experimental Researches in Chemistry and Physics* (1859). London: R. Taylor and W. Francis.

Farley, John, and Gerald Geison (1982). "Science, Politics and Spontaneous Generation in Nineteenth-Century France: The Pasteur–Pouchet Debate," pp. 1–38 in H. M. Collins (Ed.), *Sociology of Scientific Knowledge: A Source Book.* Bath: Bath University Press. [Originally, *Bulletin of the History of Medicine* (1974), **48**:161–198.]

Fauvel, John, Raymond Flood, Michael Shortlans, and Robin Wilson (Eds.) (1988). *Let Newton Be! A New Perspective on His Life and Works.* Oxford: Oxford University Press.

Fermi, Laura, and Gilberto Bernardini (1961). *Galileo and the Scientific Revolution.* New York: Basic Books.

Fisher, R. A. (1922). "On the Mathematical Foundations of Theoretical Statistics," *Philosophical Transactions the Royal Society of London,* **A222**:309–368.

——— (1926). *The Design of Experiments* (4th ed., 1947). Edinburgh: Oliver & Boyd.

——— (1935). "Statistical Tests," *Nature,* **136**:474.

——— (1936). "Has Mendel's Work Been Rediscovered?" *Annals of Science,* **1**:115–137.

——— (1956). *Statistical Methods and Statistical Inference,* Edinburgh: Oliver & Boyd.

——— (1958 [1930]). *The Genetical Theory of Natural Selection.* Oxford: Oxford University Press; New York: Dover.

——— (1970). *Statistical Methods for Research Workers,* 14th ed. Edinburgh: Oliver & Boyd (first published in 1925).

Fletcher, R. (1991). *Science, Ideology, and the Media: The Cyril Burt Scandal.* New Brunswick, NJ: Transaction.

Fliess, Wilhelm (1906). *In Eigener Sache: Gegen Otto Weininge und Hermann Swoboda.* Berlin: E. Goldschmidt.

Florens, Jean-Pierre, Michel Mouchart, and Jean-Marie Roin (1990). *Elements of Bayesian Statistics.* New York: Marcel Dekker.

Frank, Philipp (1947, 1953). *Einstein: His Life and Times,* English translation. New York: Alfred A. Knopf.

Franklin, Kenneth J. (1961). *William Harvey: Englishman.* London: Macgibbon & Kee.

Freeman, Derek (1983). *Margaret Mead and Samoa: The Making and Unmaking of an Anthropological Myth.* Cambridge, MA: Harvard University Press.

Freud, Ernst L. (Ed.) (1960). *Letters of Sigmund Freud, 1873–1939,* translated by Tania and James Stern. New York: Basic Books.

Freud, Sigmund (1974). In James Strachey et al. (Eds.), *The Standard Edition of the Complete Psychological Works of Sigmund Freud,* 24 vols. London: Hogart Press and Institute of Psychoanalysis.

Fromm, Erich (1959). *Sigmund Freud's Mission: An Analysis of His Personality and Influence.* New York: Harper.

Galton, F. (1875). "The History of Twins as a Criterion of the Relative Powers of Nature and Nurture," *Fraser's Magazine*, **12**:566–576.

Gamerman, Dani (1997). *Markov Chain Monte Carlo: Stochastic Simulation for Bayesian Inference*. Boca Raton, FL: CRC Press.

Gardner, M. (1977). "Great Fakes of Science," *Esquire*, Oct. pp. 88–92.

Gatsonis, C., J. S. Hodges, R. E. Kass, and N. D. Singpurwalla (Eds.) (1993). *Case Studies in Bayesian Statistics*, New York: Springer-Verlag.

——— (1995). *Case Studies in Bayesian Statistics*, Vol. II. New York: Springer-Verlag.

Gatsonis, C., J. S. Hodges, R. E. Kass, R. McCulloch, P. Rossi, and N. D. Singpurwalla (Eds.) (1997). *Case Studies in Bayesian Statistics*, Vol. III. New York: Springer-Verlag.

Gatsonis, C., B. Carlin, A. Gelman, M. West, R. E. Kass, A. Carriquiry, and I. Verdinelli (Eds.) (1999). *Case Studies in Bayesian Statistics*, Vol. IV. New York: Springer-Verlag.

Geison, Gerald A. (1974). "Louis Pasteur," pp. 350–416 in Charles C. Gillispie (Ed.), *The Dictionary of Scientific Biography*, Vol. 10. New York: Scribners.

——— (1995). *The Private Science of Louis Pasteur*. Princeton, NJ: Princeton University Press.

Geisser, Seymour (1993). *Predictive Inference: An Introduction*. New York: Chapman & Hall.

Gelfand, A. E., and A. F. M. Smith (1990). "Sampling Based Approaches to Calculating Marginal Densities," *Journal of the American Statistical Association*, **85**:398–409.

Gelman, Andrew, John B. Carlin, Hal S. Stern, and Donald B. Rubin (1995). *Bayesian Data Analysis*. London: Chapman & Hall.

Geman, S., and D. Geman (1986). "Stochastic Relaxation, Gibbs Distributions and the Bayesian Restoration of Images," *IEEE Transactions on Pattern Analysis and Machine Intelligence*, **6**:721–741.

Gholson, Barry, William R. Shadish, Jr., Robert A. Neimeyer, and Arthur C. Houts (Eds.) (1989). *Psychology of Science: Contributions to Metascience*. Cambridge: Cambridge University Press.

Gibson, Charles (1970). *Heroes of the Scientific World*. Freeport, NY: Books for Libraries Press.

Gilks, W. R., S. Richardson, and D. J. Spiegelhalter (Eds.) (1996). *Markov Chain Monte Carlo in Practice*. Boca Raton, FL: CRC Press.

Gillie, Oliver (1976). "Crucial Data Was Faked by Eminent Psychologist," *Sunday Times* [London], October 24, 1976, p. 1.

Gillispie, Charles Coulston (1990 [1960]). *The Edge of Objectivity: An Essay in the History of Scientific Ideas*. Princeton, NJ: Princeton University Press.

——— (Ed.) (1974). *Dictionary of Scientific Biography*. New York: Charles Scribner's Sons.

Giroud, Françoise (1986). *Marie Curie: A Life*, translated by Lydia Davis from *Une Femme Honorable*, copyright Librarie Arthème Fayard, 1981. New York: Holmes & Meier.

Glanz, James (2000). "New Tactic in Physics: Hiding the Answer," *Science Times, in The New York Times*, August. 8, 2000, pp. D1, and D4.

Glass, B. (1963). "The Establishment of Modern Genetical Theory as an Example of the Interaction of Different Models, Techniques, and Inferences," pp. 521–541 in A. C. Crombie (Ed.), *Scientific Change*. New York: Basic Books; London: Heinemann.

Golinski, Jan (1990). "The Theory of Practice and the Practice of Theory: Sociological Approaches to the History of Science," *Isis*, **81**:492–505.

Good, I. J. (1950). *Probability and the Weighting of Evidence*. London: Charles Griffin.

——— (1965). *The Estimation of Probabilities: An Essay on Modern Bayesian Methods*, Research Monograph 30. Cambridge, MA: MIT Press.

——— (1974). *Information, Weight of Evidence, the Singularity Between Probability Measures and Signal Detection*. New York: Springer-Verlag.

——— (1983). *Good Thinking: The Foundations of Probability and Its Applications*. Minneapolis, MN: University of Minnesota Press.

Gooding, David (1985). "'In Nature's School': Faraday as an Experimentalist," pp. 105–135 in David Gooding and Frank A. J. L. James (Eds.), *Faraday Rediscovered: Essays on the Life and Work of Michael Faraday, 1791–1867*. Bassingstoke, Hants, UK: Macmillan. (Published in the United States and Canada by Stockton Press, New York.)

Gooding, David, and Frank A. L. J. James (Eds.) (1985). *Faraday Rediscovered: Essays on the Life and Work of Michael Faraday, 1791–1867*. Bassingstoke, Hants, UK: Macmillan. (Published in the United States and Canada by Stockton Press, New York, NY.)

Gould, Stephen Jay (1978). "The Finagle Factor," *Human Nature*, July, pp. 294–299.

——— (1981). *The Mismeasure of Man*. New York: W.W. Norton.

Grant, Sir Alexander (1877). *Aristotle*. Edinburgh: William Blackwood and Sons.

Gray, P. (1983). "Fakes That Have Skewed History," *Newsweek*, May 16, p. 25.

Gray, Paul (1993). "The Assault on Freud," *Time*, November 29, pp. 47–51.

Grollman, Earl A. (1965). *Judaism in Sigmund Freud's World*. New York: Bloch Publishing Co.

Gruber, Howard E., and Paul H. Barrett (1974). *Darwin on Man*. Toronto: Clarke, Irwin.

Grunbaum, Adolf (1983). In R. S. Cohen and L. Laudan (Eds.), *Physics, Philosophy, and Psychoanalysis: Essays in Honor of Adolf Grünbaum*. Dordrecht, The Netherlands: D. Reidel (distributed in the United States by Kluwer Boston).

——— (1993). *Validation in the Clinical Theory of Psychoanalysis: A Study in the Philosophy of Psychoanalysis*. Madison, CT: International Universities Press.

Guerlac, Henry (1961). *Lavoisier—The Crucial Year: The Background and Origin of His First Experiments on Combustion in 1772*. Ithaca, NY: Cornell University Press.

——— (1973). "Lavoisier," in Charles Coulston Gillispie (Ed.), *Dictionary of Scientific Biography*. New York: Charles Scribner's Sons.

Hacking, Ian (1965). *Logic of Scientific Inference*, Cambridge: Cambridge University Press.

Hadamard, Jacques (1945). *An Essay on the Psychology of Invention in the Mathematical Field*. Princeton, NJ: Princeton University Press.

Hamilton, John (1991). *They Made Our World: Five Centuries of Great Scientists and Inventors*. London: Broadside Books.

Hartigan, J. A. (1983). *Bayes Theory*. New York: Springer-Verlag.

Hearnshaw, L. S. (1979). *Cyril Burt: Psychologist*, London: Hodder & Stoughton.

Herrnstein, Richard J., and Charles Murray (1994). *The Bell Curve: Intelligence and Class Structure in American Life*. New York: Free Press.

Hetherington, N. S. (1983). "Just How Objective Is Science?" *Nature*, **306**:727.

Higgins, A. C. (1995). *Science Fraud Bulletin Board* (Internet address: scifraud@albnyvm1), June 12, 13, December 26.

———(1994). *Bibliography on Scientific Fraud*. Albany, NY: Exams Unlimited.

Hoffman, Banesh (with the collaboration of Helen Dukas) (1972). *Albert Einstein: Creator and Rebel*. New York: Viking.

Holmes, Frederic Lawrence (1985). *Lavoisier and the Chemistry of Life: An Exploration of Scientific Creativity*. Madison, WI: University of Wisconsin Press.

———(1987). "Scientific Writing and Scientific Discovery," *Isis*, **78**:220–235.

———(1988). "Lavoisier's Conceptual Passage," *Osiris*, **4**:82–92.

———(1990). "Laboratory Notebooks: Can the Daily Record Illuminate the Broader Picture?" *Proceedings of the American Philosophical Society*, **134**:349–366.

Holton, Gerald (1978). "Subelectrons, Presuppositions, and the Millikan–Ehrenhaft Dispute," pp. 25–83 in Gerald Holton, *The Scientific Imagination: Case Studies*. Cambridge: Cambridge University Press.

———(1993). *Science and Anti-science*. Cambridge, MA: Harvard University Press.

———(1996). *Einstein, History and Other Passions*. Reading, MA: Addison-Wesley.

Howson, Colin, and Peter Urbach (1989). *Scientific Reasoning: The Bayesian Approach*. La Salle, IL: Open Court.

Hubbard, Ruth (1997). "Irreplaceable Ewe," Editorial, *The Nation*, February 24.

Hulbert, Mark (1999). "Are Fund Managers Irrelevant? An 18th Century Theory Suggests Not," *New York Times*, October 10, Business Section, p. 26.

Huxley, Sir Julian (1958). "Introduction," mentor edition of *The Origin of Species by Charles Darwin*. New York: Mentor Books.

Jaffe, Aniela (Ed.) (1963). *Carl Jung: Memories, Dreams, Reflections*, translated from the German. New York: Vintage Books.

Jastrow, Joseph (1932). *Freud: His Dream and Sex Theories*. New York: Pocket Books.

Jaynes, E. T. (1983). In R. D. Rosenkrantz (Ed.), *Papers on Probability, Statistics, and Statistical Physics of E. T. Jaynes*. Dordrecht, The Netherlands: D. Reidel.

Jeffreys, Harold (1939, 1948, 1961). *Theory of Probability*, 1st, 2nd, and 3rd editions. Oxford: Oxford University Press.

Jensen, A. R. (1969). "How Much Can We Boost IQ and Scholastic Achievement?" *Harvard Educational Review*, **39**:1–123.

———(1995). "IQ and Science: The Mysterious Burt Affair," in N. J. Mackintosh (Ed.), *Cyril Burt: Fraud or Framed*. Oxford: Oxford University Press.

Johnson, George (1997). "Don't Worry. A Brain Still Can't Be Cloned," Week in Review, *New York Times*, March 2, p. 1.

Jones, Bence (1870). *The Life and Letters of Faraday*, 2 vols. London: Longmans, Green.

Jones, Ernest (1953–1957). *The Life and Works of Sigmund Freud*, 3 vols. New York: Basic Books.

Joynson, Robert B. (1989). *The Burt Affair*. New York: Routledge.

Kadane, Joseph B., Mark J. Schervish, and Teddy Seidenfeld (1999). *Rethinking the Foundations of Statistics*. Cambridge: Cambridge University Press.

Kamin, L. J. (1974). *The Science and Politics of IQ*. New York: John Wiley & Sons.

Keller, Evelyn Fox (1985). *Reflections on Gender and Science*. New Haven, CT: Yale University Press.

Kerr, John (1950). *A Most Dangerous Method: The Story of Jung, Freud, and Sabina Spielrein*. New York: Alfred A. Knopf, distributed by Random House.

Keynes, Geoffrey (1966). *The Life of William Harvey*. Oxford: Oxford University Press.

Keynes, John Maynard (1921). *A Treatise on Probability*. London: Macmillan.

Klencke, H. (1852). *Alexander von Humboldt: A Biographical Monument*, translated by Juliette Bauer. London: Ingram, Cooke.

Kline, Paul (1972). *Fact and Fancy in Freudian Theory*. London: Methuen.

Koestler, Arthur (1973). *The Case of the Midwife Toad*. New York: Vintage Books.

Kohn, Alexander (1997). *False Prophets: Fraud and Error in Science and Medicine*, rev. ed. New York: Barnes & Noble.

Koyre, Alexander A. (1968). *Metaphysics and Measurement*. Cambridge, MA: Harvard University Press.

Kuhn, Thomas S. (1962). *The Structure of Scientific Revolutions*, 2nd ed. enlarged. Chicago: University of Chicago Press.

Lad, Frank (1996). *Operational Subjective Statistical Methods: A Mathematical, Philosophical, and Historical Introduction*. New York: Wiley-Interscience.

Lakatos, I. (1968). "Criticism and Methodology of Scientific Research Programmes," *Proceedings of the Aristotelian Society*, **69**:149–186.

——— (1970). "Falsification and the Methodology of Scientific Research Programmes," in I. Lakatos and A. Musgrave (Eds.), *Criticism and the Growth of Knowledge*. Cambridge: Cambridge University Press.

——— (1974). "Popper on Demarcation and Induction," Chapter 5 in P. A. Schilpp (Ed.), *The Philosophy of Karl Popper*, Vol. 2. LaSalle, IL: Open Court.

——— (1978). In J. Worrall, and G. Currie (Eds.), *Philosophical Papers*, 2 vols. Cambridge: Cambridge University Press.

Laplace, P. S. (1774). "Mémoire sur la probabilité des causes par les événements," *Memoires de l'Academie Royale des Sciences Presents Par Divers Savans*, **6**:621–656.

Lee, Peter M. (1997). *Bayesian Statistics: An Introduction*, 2nd ed. New York: Wiley.

Lenard, Philipp Eduard Anton (1933). *Great Men of Science: A History of Scientific Progress*. New York: Macmillan.

Leonard, Thomas, and John S. J. Hsu (1999). *Bayesian Methods: An Analysis for Statisticians and Interdisciplinary Researchers*. Cambridge: Cambridge University Press.

Lindley, Dennis V. (1965). *Introduction to Probability and Statistics*, 2 vols. (Part 1: Probability; Part 2: Inference). Cambridge: Cambridge University Press.

———(1972). *Bayesian Statistics: A Review*. Philadelpia: Society for Industrial and Applied Mathematics.

Lloyd, G. E. R. (1987). "Empirical Research in Aristotle's Biology," pp. 59ff. in Allan Gotthelf and James Lennox (Eds.), *Philosophical Issues in Aristotle's Biology*. Cambridge: Cambridge University Press.

Loir, Adrien (1937–1938). "A l'ombre de Pasteur" (In the Shadow of Pasteur), *Mouvement Sanitaire*, **14**:91–92.

Longino, Helen E., E. F. Keller, Anne Fausto-Sterling, and Sandra Harding (1988). "Science, Objectivity, and Feminist Values: Review Essay," *Feminist Studies*, **14**:561–574.

MacDonald, D. K. C. (1964). *Faraday, Maxwell, and Kelvin*. Garden City, NY: Doubleday.

Mach, Ernst (1960). *The Science of Mechanics*. La Salle, IL: Open Court.

Mackintosh, N. J. (Ed.) (1995). *Cyril Burt: Fraud or Framed?* Oxford: Oxford University Press.

Mahoney, Michael J. (1979). "Psychology of the Scientist: An Evaluative Review," *Social Studies of Science*, **9**:349–375.

Malakoff, David (1999). "Bayes Offers a 'New' Way to Make Sense of Numbers," *Science*, **286**:1460–1464.

Mathews, Jon, and R. L. Walker (1965). *Mathematical Methods of Physics*. New York: W. A. Benjamin, Inc.

Matthews, Robert A. J. (1998a). *Fact Versus Factions: The Use and Abuse of Subjectivity in Scientific Research*, ESEF Working Paper 2/98, European Science and Environment Forum, Cambridge, September.

———(1998b). "Flukes and Flaws," *Prospect*, Issue 35, pp. 20–24.

McKenzie, A. E. E. (1960). *The Major Achievements of Science: The Development of Science from Ancient Times to the Present*. New York: Simon & Schuster.

McKie, Douglas (1952). *Antoine Lavoisier: Scientist, Economist, Social Reformer*. London: Constable and Co.; New York: Henry Schuman.

Mead, Margaret (1973 [1928]). *Coming of Age in Samoa*. New York: Morrow.

Meadows, Jack (1987). *The Great Scientists*. New York: Oxford University Press.

Medawar, Peter B. (1984). *Pluto's Republic*. Oxford, UK: Oxford University Press.

———(1964). "Is the Scientific Paper Fraudulent," *Saturday Review*, August 1, 1964, pp. 42–43.

Meeden, G., and Malay Ghosh (1997). *Bayesian Methods for Finite Population Sampling*. Boca Raton, FL: CRC Press.

Mendel Centennial Symposium (1965: Fort Collins (Colo.) (1967). *Heritage from Mendel: Proceedings of the Mendel Centennial Symposium Sponsored by the Genetics Society of America*. Madison, WI: Wisconsin University Press.

Mendel, Gregor (1965 [1866]). *Experiments in Plant Hybridization*, Cambridge, MA: Harvard University Press.

Metchnikoff, Elie (1933). *Trois Fondateurs de la Médecine Moderne: Pasteur, Lister, Koch*. Paris: Librairie Felix Alcan.

Michelmore, Peter (1962). *Einstein: Profile of the Man*. New York: Dodd.

Millar, David (1989). *Chambers Concise Dictionary of Scientists*. Cambridge: W & R Chambers and the Press Syndicate of the University of Cambridge.

Miller, Arthur I. (1996). *Insights of Genius*. New York: Springer-Verlag.

Miller, Johnathan, and Borin van Loon (1982). *Darwin for Beginners*. New York: Pantheon Books.

Miller, Mabel (1968). *Michael Faraday and the Dynamo*. Philadelphia: Chilton.

Millikan, Robert A. (1909–1910). "A New Modification of the Cloud Method of Measuring the Elementary Electrical Charge, and the Most Probable Value of that Charge," *Physical Review*, **29**:560–561 (Abstract, 1909); the full paper was published in 1910 in *Philosophical Magazine*, pp. 209–228.

———(1913). "On the Elementary Electrical Charge and the Avogadro Constant," *Physical Review*, **2**:109–143.

Mockus, Jonas, William Eddy, Audris Mockus, Linas Mockus, and G. V. Reklaitis (1997). *Bayesian Heuristic Approach to Discrete and Global Optimization: Algorithms, Visualization, Software, and Applications*. Boston: Kluwer Academic Publishers.

Monaghan, F., and A. Corcos (1985a). Chi-Square and Mendel's Experiments: Where's the Bias? *Journal of Heredity*, **76**:307–309.

———(1985b). "Mendel, the Empiricist," *Journal of Heredity*, **76**:49–54.

Mosteller, Frederick, and David L. Wallace (1984). *Applied Bayesian and Classical Inference*, 2nd ed. New York: Springer-Verlag.

Mulkay, Michael J., and Gilbert, G. Nigel (1982). "Accounting for Error: How Scientists Construct Their Social World When They Account for Correct and Incorrect Belief," *Sociology*, **16**:165–183.

Muller, Peter, and Brani Vidakovic (Eds.) (1999). *Bayesian Inference in Wavelet-Based Models*, Vol. 141 of Lecture Notes in Statistics. New York: Springer-Verlag.

Neal, Radford M. (1996). *Bayesian Learning for Neural Networks*, Vol. 118 of Lecture Notes in Statistics. New York: Springer-Verlag.

Nelken, Halina (1980). *Alexander von Humboldt: His Portraits and Their Artists: A Documentary Iconography*. Berlin: D. Reimer.

Nelkin, Dorothy (1995). *Selling Science: How the Press Covers Science and Technology*, rev. ed. New York: W.H. Freeman.

Neyman, Jerzy, and Lucian M. Le Cam (Eds.) (1965). *Bernoulli (1713), Bayes (1763), Laplace (1813)*, Anniversary Volume. New York: Springer-Verlag.

Neyman, J., and E. S. Pearson (1928). "On the Use and the Interpretation of Certain Test Criteria for Purposes of Statistical Inference," *Biometrika*, **20**:175–240, Part I; 263–294, Part II.

———(1933). "On the Problem of the Most Efficient Tests of Statistical Hypotheses," *Philosophical Transactions of the Royal Society*, **231A**:289–337.

Nicolle, Jacques (1953). *Un Maître de l'Enquete Scientifique, Louis Pasteur*. Paris: La Colombe. Translated as *Louis Pasteur: A Master of Scientific Enquiry*. London: Hutchinson, 1961.

———(1966). *Louis Pasteur: The Story of His Major Discoveries*. New York: Fawcett.

Ogle, W. (Trans.) (1897). *Aristotle on Youth and Old Age, Life and Death and Respiration*. London: Longmans.

O'Hagen, Anthony (1994). "Bayesian Inference," in *Kendall's Advanced Theory of Statistics*, Vol. 2B. London: Edward Arnold/Hodder Headline.

Olby, Robert C. (1966). *Origins of Mendelism*, New York: Schocken Books.

Orans, Martin (1996). *Not Even Wrong: Margaret Mead, Derek Freeman, and the Samoans*. Novato, CA: Chandler & Sharp.

Orel, V. (1968). "Will the Story on 'Too Good' Results of Mendel's Data Continue?" *Bioscience*, **18**:776–778.

O'Ruanaidh, Joseph J. K., and William J. Fitzgerald (1996). *Numerical Bayesian Methods Applied to Signal Processing*. New York: Springer-Verlag.

Pagel, Walter (1967). *William Harvey's Biological Ideas: Selected Aspects and Historical Background*. New York: Hafner.

———(1976). *New Light on William Harvey*. New York: S. Karger.

Perutz, M. F. (1995). "The Pioneer Defended," a review of Gerald L. Geison, *The Private Science of Louis Pasteur* (Princeton, NJ: Princeton University Press), *New York Review of Books*, December 21, pp. 54ff.

Peters, Heinz E. (1962). *My Sister, My Spouse: A Biography of Lou Andreas-Salome*. New York: W.W. Norton.

Pflaum, Rosalynd (1989). *Grand Obsession*. New York: Doubleday.

Polanyi, Michael (1958, 1962). *Personal Knowledge: Towards a Post-Critical Philosophy*. Chicago: University of Chicago Press (corrected edition, 1962).

Pollard, William E. (1986). *Bayesian Statistics for Evaluation Research: An Introduction*. Beverly Hills, CA: Sage Publications.

Popper, K. R. (1956, 1982). *Quantum Theory and the Schism in Physics*, from the Postscript to *The Logic of Scientific Discovery*, W. W. Bartley III (Ed.). Totowa, NJ: Rowman & Littlefield.

———(1959a). "The Propensity Interpretation of Probability," *British Journal for the Philosophy of Science*, **10**:25–42.

———(1959b). *The Logic of Scientific Discovery*. London: Hutchinson; New York: Harper & Row.

———(1963). *Conjectures and Refutations*. New York: Routledge and & Kegan.

Posin, Dan Q. (1961). *Dr. Posin's Giants: Men of Science*. Evanston, IL: Row, Peterson.

Prebys, Eric (2000). Personal communication, August 23, 2000 and August 27, 2000.

Press, S. James (1989). *Bayesian Statistics: Principles, Models and Applications*. New York: John Wiley & Sons.

Puner, Helen W. (1947). *Freud: His Life and His Mind, A Biography*. New York: Howell, Soskin.

Quinn, Susan (1995). *Marie Curie*. New York: Simon & Schuster.

Raiffa, Howard, and Robert Schlaifer (1961). *Applied Statistical Decision Theory*. Boston: Harvard Graduate School of Business Administration.

Ramsey, Frank Plimpton (1926). "Truth and Probability," in R. B. Braithwaite (Ed.), *The Foundations of Mathematics and Other Logical Essays*. New York: Humanities Press, 1950. Reprinted in Henry E. Kyburg and Howard F. Smokler (Eds.), *Studies in Subjective Probability*, 1980. Huntington, NY: Robert E. Krieger, pp. 23–52.

Redondi, Pietro (1987). *Galileo: Heretic*, translated by Raymond Rosenthal. Princeton, NJ: Princeton University Press.

Regis, Edward (1987). *Who Got Einstein's Office*. Reading, MA: Addison-Wesley.

Reik, Theodor (1940). *From Thirty Years with Freud*. London: Hogarth Press.

Rensberger, Boyce (1983). "Margaret Mead: The Nature–Nurture Debate I," *Science83*: 28–37.

Reston, James, Jr. (1994). *Galileo: A Life*. New York: HarperCollins.

Robert, Marthe (1964). *La Revolution Psychoanalytique: La Vie et l'Oeuvre de Sigmund Freud*, 2 vols.; English translation, *The Psychoanalytic Revolution: Sigmund Freud's Life and Achievement* (1966). New York: Harcourt, Brace and World.

Roberts, Royston M. (1989). *Serendipity: Accidental Discoveries in Science*. New York: John Wiley & Sons.

Rosenthal, Robert, and Lenore Jacobson (1968). *Pygmalion in the Classroom: Teacher Expectation and Pupils' Intellectual Development*. New York: Holt, Rinehart and Winston.

Ross, Andrew (Ed.) (1996). *Science Wars*. Durham, NC: Duke University Press.

Rostand, J. (1960). *Error and Deception in Science*. New York: Basic Books.

Roux, Emle (1896). "L'Oeuvre Médicale de Pasteur," in *Agenda du Chimiste*. Paris. Reprinted in *Institut Pasteur*, "Pasteur 1822–1922." Translated by E. F. Smith as "The Medical Work of Pasteur," *Scientific Monthly*, **21**:325–389 (1925).

Rubin, Donald B. (1979). "Burt's Tables," Letters to the Editor, *Science*, **204**:245–246.

Rubin, Donald B., and Stephen M. Stigler (1979). "Dorfman's Data Analysis," Letters to the Editor, *Science*, **205**:1204–1206.

Russell, B. (1959). *The Wisdom of the West*. New York: Doubleday.

Ryan, Dennis P. (Ed.) (1987). *Einstein and the Humanities*. New York: Greenwood Press.

Sachs, Hanns (1944). *Freud: Master and Friend*. Cambridge, MA: Harvard University Press.

Sander, P., and R. Badoux (Eds.) (1991). *Bayesian Methods in Reliability*. Boston: Kluwer Academic Publishers.

Savage, L. J. (1954). *The Foundations of Statistics*. New York: John Wiley & Sons.

———(1981). *The Writings of Leonard Jimmie Savage—A Memorial Selection*. Washington, DC: American Statistical Association and Institute of Mathematical Statistics.

Schaffer, Simon (1989). "Glass Works: Newton's Prisms and the Uses of Experiment," in David Gooding, Trevor Pinch, and Simon Schaffer (Eds.), *The Uses of Experiment: Studies in the Natural Sciences*. Cambridge: Cambridge University Press.

Scheffler, Israel (1967). *Science and Subjectivity*. Indianapolis, IN: Bobbs-Merrill.

Schilpp, Paul A. (Ed.) (1949). *Albert Einstein: Philosopher–Scientist*, 2nd ed. Evanston, IL: Library of Living Philosophers.

Schlaifer, Robert (1961). *Introduction to Statistics for Business Decisions*. New York: McGraw-Hill.

Schmitt, Samuel A. (1969). *Measuring Uncertainty: An Elementary Introduction to Bayesian Statistics*. Reading, MA: Addison-Wesley.

Science Digest (1954). *Science Milestones: The Story of the Epic Scientific Achievements and the Men Who Made Them Possible*. Chicago: Windsor Press.

Seelig, Carl (1956). *Albert Einstein: A Documentary Biography*. London: Staples Press.

Seidenfeld, Teddy (1998). "P's in a Pod: Some Recipes for Cooking Mendel's Data." Pittsburgh, PA: Departments of Philosophy and Statistics, Carnegie Mellon University. Mimeo.

Shapere, Dudley (1974). *Galileo: A Philosophical Study.* Chicago: University of Chicago Press.

Shapin, Steven (1982). "History of Science and Its Sociological Reconstructions," *History of Science,* **XX**:157–211.

Shweder, Richard A. (1995). "It's Time to Reinvent Freud," Op Ed., *New York Times,* December 15, p. A19.

Simmons, John (1996). *The Scientific 100: A Ranking of the Most Influential Scientists, Past and Present.* Secaucus, NJ: Carol Publishing Group, Citadel Press.

Singer, Charles (1956). *The Discovery of the Circulation of the Blood.* London: Wm. Dawson & Sons.

Smith, C. Ray (Ed.) (1985). *Maximum-Entropy and Bayesian Methods in Inverse Problems,* Vol. 14 of (Fundamental Theories of Physics). Dordrecht, The Netherlands: D. Reidel.

Solmsen, Friedrich (1960). *Aristotle's System of the Physical World: A Comparison With His Predecessors,* Ithaca, NY: Cornell University Press.

Specter, Michael, with Gina Kolata (1997). "After Decades and Many Missteps, Cloning Success," *New York Times,* March 3, p. 1.

Stearn, William T. (Ed.) (1968). *Humboldt, Bonpland, Kunth and Tropical American Botany: A Miscellany on the "Nova Genera et Species Plantarum."* Stuttgart, Germany: Verlag von J. Carmer.

Stekel, Wilhelm (1950). *Autobiography: The Life Story of a Pioneer Psychoanalyst.* New York: Liveright.

Stigler, Stephen M. (1979). "Burt's Tables," Letters to the Editor, *Science,* **204**:242–245.

———(1986). *The History of Statistics.* Cambridge, MA: Harvard University Press.

Stone, Lawrence D., Carl A. Barlow, and Thomas L. Corwin (1999). *Bayesian Multiple Target Tracking.* Norwood, MA: Artech House.

Sturtevant, A. H. (1965). *A History of Genetics.* Tokyo: Weatherhill.

Susac, Andrew (1970). *The Clock, the Balance, and the Guillotine: The Life of Antoine Lavoisier.* Garden City, NY: Doubleday.

Tanner, Martin A. (1996). *Tools for Statistical Inference: Methods for the Exploration of Posterior Distributions and Likelihood Functions,* 3rd ed. New York: Springer-Verlag.

Thomas, John Meurig (1991). *Michael Faraday and the Royal Institution (The Genius of Man and Place).* Bristol, UK: Adam Hilger.

Torrey, E. Fuller (1992). *Freudian Fraud: The Malignant Effect of Freud's Theory on American Thought and Culture.* New York: HarperCollins.

Tricker, R. A. R. (1966). *The Contributions of Faraday and Maxwell to Electrical Science.* Oxford: Pergamon Press.

Trilling, Lionel, and Steven Marcus (1961). *The Life and Works of Sigmund Freud,* edited and abridged. New York: Penguin Books.

Vallentin, Antonina (1954). *Einstein: A Biography*, English translation. Garden City, NY: Doubleday; Paris: Plon et les Librairies Associés.

Vallery-Radot, René (1883). *Pasteur: Histoire d'un Savant par un Ignorant*. Paris: J. Hetzel.

———(1900). *La Vie de Pasteur*, 2 vols. Paris: Flammarion. Translated as *The Life of Pasteur* by R. L. Devonshire, 2 vols. London: Constable, 1911.

Van Der Waerden, B. L. (1968). "Mendel's Experiments," *Centaurus*, **12**:275–288.

Van Dusen, Robert (1971). *The Literary Ambitions and Achievements of Alexander von Humboldt*. Berne, Switzerland: Herbert Lang; Frankfurt am Main, Germany: Peter Lang.

Veatch, Henry B. (1974). *Aristotle: A Contemporary Appreciation*. Bloomington, IN: Indiana University Press.

Viertl, R. (Ed.) (1987). *Probability and Bayesian Statistics*. New York: Plenum Press.

von Hagen, Victor Wolfgang (1945). *South America Called Them: Explorations of the Great Naturalists: La Condamine, Humboldt, Darwin, Spruce*. New York: Alfred A. Knopf.

Von Tschermak, E. (1900). Uber künstliche Kreuaung bei *Pisum Sativum*," *Botanischen Gesellschaft Berichte*, **18**:232–239.

Watson, James D., and Francis H. C. Crick (1953). "A Structure for Deoxyribose Nucleic Acid." *Nature*, **171**:737.

Weber, Max (1946). "Science and the Disenchantment of the World," in H. H. Gerth, and C. Wright Mills (Eds.), *From Max Weber: Essays in Sociology*. Oxford: Oxford University Press.

Weiling, F. (1971). "Mendel's Too Good Data," in Special Issue on *Pisum* Experiments, *Folia Mendelania*, **6**:71–77.

West, Mike, and Jeff Harrison (1997). *Bayesian Forecasting and Dynamic Models*. New York: Springer-Verlag.

Westfall, Richard (1973). "Newton and the Fudge Factor," *Science*, **179**:751–758.

Whitteridge, Gweneth (1967). "William Harvey," pp. 673–675, Vol. 11, *Collier's Encyclopedia*. New York: Crowell Collier Macmillan.

———(1971). *William Harvey and the Circulation of the Blood*. London: Macdonald; New York: American Elsevier.

Williams, L. Pearce (1965). *Michael Faraday: A Biography*. London: Chapman & Hall.

———(1966). *The Origins of Field Theory*. New York: Random House.

Wilmut, I., A. E. Schnieke, J. McWhir, A. J. Kind, and K. H. S. Campbell (1997). "Viable Offspring Derived from Fetal and Adult Mammalian Cells," Letter, *Nature*, **385**:810.

Wilson, Curtis (1968). "Kepler's Derivation of the Elliptical Path," *Isis*, **59**:75–90.

Wilson, Grove (1937). *Great Men of Science*. Garden City, NY: Garden City Publishing Company.

Winkler, Robert L. (1972). *An Introduction to Bayesian Inference and Decision*. New York: Holt, Rinehart and Winston.

Wittels, Fritz (1924). *Sigmund Freud*, English translation. Leipzig: E.P. Tal.

Wolpert, Lewis (1992). *The Unnatural Nature of Science*. Boston: Harvard University Press.

———(1995). "Experiments in Deceit: A Review of Gerald L. Geison, *The Private Science of Louis Pasteur,*" Book Review, *New York Times,* May 7, 1995, p. 35.

Wortis, Joseph (1954). *Fragments of an Analysis with Freud.* New York: Simon & Schuster.

Wright, Lawrence (1997). *Twins: And What They Tell Us About Who We Are.* New York: John Wiley & Sons.

Wright, S. (1966). "Mendel's Ratios," in C. Stern and E. R. Sherwood (Eds.), *The Origin of Genetics: A Mendel Source Book.* San Francisco: W.H. Freeman.

Wyatt, R. B. Hervey (1924). *William Harvey.* London: Beonard Parsons; Boston: Small, Maynard.

Zellner, Arnold (1971). *An Introduction to Bayesian Inference in Econometrics.* New York: John Wiley & Sons.

———(1997). *Bayesian Analysis in Econometrics and Statistics: The Zellner View and Papers.* Cheltenham, UK: Edward Elgar.

Zinder, N., and A. Meyer (1983). "Fraud in Science: A Scientist's View," *Science,* **83**:94.

Subject Index

Name Index

A CATALOG OF SELECTED
DOVER BOOKS
IN SCIENCE AND MATHEMATICS

Astronomy

CHARIOTS FOR APOLLO: The NASA History of Manned Lunar Spacecraft to 1969, Courtney G. Brooks, James M. Grimwood, and Loyd S. Swenson, Jr. This illustrated history by a trio of experts is the definitive reference on the Apollo spacecraft and lunar modules. It traces the vehicles' design, development, and operation in space. More than 100 photographs and illustrations. 576pp. 6 3/4 x 9 1/4. 0-486-46756-2

EXPLORING THE MOON THROUGH BINOCULARS AND SMALL TELESCOPES, Ernest H. Cherrington, Jr. Informative, profusely illustrated guide to locating and identifying craters, rills, seas, mountains, other lunar features. Newly revised and updated with special section of new photos. Over 100 photos and diagrams. 240pp. 8 1/4 x 11. 0-486-24491-1

WHERE NO MAN HAS GONE BEFORE: A History of NASA's Apollo Lunar Expeditions, William David Compton. Introduction by Paul Dickson. This official NASA history traces behind-the-scenes conflicts and cooperation between scientists and engineers. The first half concerns preparations for the Moon landings, and the second half documents the flights that followed Apollo 11. 1989 edition. 432pp. 7 x 10.

0-486-47888-2

APOLLO EXPEDITIONS TO THE MOON: The NASA History, Edited by Edgar M. Cortright. Official NASA publication marks the 40th anniversary of the first lunar landing and features essays by project participants recalling engineering and administrative challenges. Accessible, jargon-free accounts, highlighted by numerous illustrations. 336pp. 8 3/8 x 10 7/8. 0-486-47175-6

ON MARS: Exploration of the Red Planet, 1958-1978--The NASA History, Edward Clinton Ezell and Linda Neuman Ezell. NASA's official history chronicles the start of our explorations of our planetary neighbor. It recounts cooperation among government, industry, and academia, and it features dozens of photos from Viking cameras. 560pp. 6 3/4 x 9 1/4. 0-486-46757-0

ARISTARCHUS OF SAMOS: The Ancient Copernicus, Sir Thomas Heath. Heath's history of astronomy ranges from Homer and Hesiod to Aristarchus and includes quotes from numerous thinkers, compilers, and scholasticists from Thales and Anaximander through Pythagoras, Plato, Aristotle, and Heraclides. 34 figures. 448pp. 5 3/8 x 8 1/2.

0-486-43886-4

AN INTRODUCTION TO CELESTIAL MECHANICS, Forest Ray Moulton. Classic text still unsurpassed in presentation of fundamental principles. Covers rectilinear motion, central forces, problems of two and three bodies, much more. Includes over 200 problems, some with answers. 437pp. 5 3/8 x 8 1/2. 0-486-64687-4

BEYOND THE ATMOSPHERE: Early Years of Space Science, Homer E. Newell. This exciting survey is the work of a top NASA administrator who chronicles technological advances, the relationship of space science to general science, and the space program's social, political, and economic contexts. 528pp. 6 3/4 x 9 1/4.

0-486-47464-X

STAR LORE: Myths, Legends, and Facts, William Tyler Olcott. Captivating retellings of the origins and histories of ancient star groups include Pegasus, Ursa Major, Pleiades, signs of the zodiac, and other constellations. "Classic." – *Sky & Telescope.* 58 illustrations. 544pp. 5 3/8 x 8 1/2. 0-486-43581-4

A COMPLETE MANUAL OF AMATEUR ASTRONOMY: Tools and Techniques for Astronomical Observations, P. Clay Sherrod with Thomas L. Koed. Concise, highly readable book discusses the selection, set-up, and maintenance of a telescope; amateur studies of the sun; lunar topography and occultations; and more. 124 figures. 26 halftones. 37 tables. 335pp. 6 1/2 x 9 1/4. 0-486-42820-6

Browse over 9,000 books at www.doverpublications.com

Chemistry

MOLECULAR COLLISION THEORY, M. S. Child. This high-level monograph offers an analytical treatment of classical scattering by a central force, quantum scattering by a central force, elastic scattering phase shifts, and semi-classical elastic scattering. 1974 edition. 310pp. 5 3/8 x 8 1/2. 0-486-69437-2

HANDBOOK OF COMPUTATIONAL QUANTUM CHEMISTRY, David B. Cook. This comprehensive text provides upper-level undergraduates and graduate students with an accessible introduction to the implementation of quantum ideas in molecular modeling, exploring practical applications alongside theoretical explanations. 1998 edition. 832pp. 5 3/8 x 8 1/2. 0-486-44307-8

RADIOACTIVE SUBSTANCES, Marie Curie. The celebrated scientist's thesis, which directly preceded her 1903 Nobel Prize, discusses establishing atomic character of radioactivity; extraction from pitchblende of polonium and radium; isolation of pure radium chloride; more. 96pp. 5 3/8 x 8 1/2. 0-486-42550-9

CHEMICAL MAGIC, Leonard A. Ford. Classic guide provides intriguing entertainment while elucidating sound scientific principles, with more than 100 unusual stunts: cold fire, dust explosions, a nylon rope trick, a disappearing beaker, much more. 128pp. 5 3/8 x 8 1/2. 0-486-67628-5

ALCHEMY, E. J. Holmyard. Classic study by noted authority covers 2,000 years of alchemical history: religious, mystical overtones; apparatus; signs, symbols, and secret terms; advent of scientific method, much more. Illustrated. 320pp. 5 3/8 x 8 1/2. 0-486-26298-7

CHEMICAL KINETICS AND REACTION DYNAMICS, Paul L. Houston. This text teaches the principles underlying modern chemical kinetics in a clear, direct fashion, using several examples to enhance basic understanding. Solutions to selected problems. 2001 edition. 352pp. 8 3/8 x 11. 0-486-45334-0

PROBLEMS AND SOLUTIONS IN QUANTUM CHEMISTRY AND PHYSICS, Charles S. Johnson and Lee G. Pedersen. Unusually varied problems, with detailed solutions, cover of quantum mechanics, wave mechanics, angular momentum, molecular spectroscopy, scattering theory, more. 280 problems, plus 139 supplementary exercises. 430pp. 6 1/2 x 9 1/4. 0-486-65236-X

ELEMENTS OF CHEMISTRY, Antoine Lavoisier. Monumental classic by the founder of modern chemistry features first explicit statement of law of conservation of matter in chemical change, and more. Facsimile reprint of original (1790) Kerr translation. 539pp. 5 3/8 x 8 1/2. 0-486-64624-6

MAGNETISM AND TRANSITION METAL COMPLEXES, F. E. Mabbs and D. J. Machin. A detailed view of the calculation methods involved in the magnetic properties of transition metal complexes, this volume offers sufficient background for original work in the field. 1973 edition. 240pp. 5 3/8 x 8 1/2. 0-486-46284-6

GENERAL CHEMISTRY, Linus Pauling. Revised third edition of classic first-year text by Nobel laureate. Atomic and molecular structure, quantum mechanics, statistical mechanics, thermodynamics correlated with descriptive chemistry. Problems. 992pp. 5 3/8 x 8 1/2. 0-486-65622-5

ELECTROLYTE SOLUTIONS: Second Revised Edition, R. A. Robinson and R. H. Stokes. Classic text deals primarily with measurement, interpretation of conductance, chemical potential, and diffusion in electrolyte solutions. Detailed theoretical interpretations, plus extensive tables of thermodynamic and transport properties. 1970 edition. 590pp. 5 3/8 x 8 1/2. 0-486-42225-9

Browse over 9,000 books at www.doverpublications.com

Engineering

FUNDAMENTALS OF ASTRODYNAMICS, Roger R. Bate, Donald D. Mueller, and Jerry E. White. Teaching text developed by U.S. Air Force Academy develops the basic two-body and n-body equations of motion; orbit determination; classical orbital elements, coordinate transformations; differential correction; more. 1971 edition. 455pp. 5 3/8 x 8 1/2. 0-486-60061-0

INTRODUCTION TO CONTINUUM MECHANICS FOR ENGINEERS: Revised Edition, Ray M. Bowen. This self-contained text introduces classical continuum models within a modern framework. Its numerous exercises illustrate the governing principles, linearizations, and other approximations that constitute classical continuum models. 2007 edition. 320pp. 6 1/8 x 9 1/4. 0-486-47460-7

ENGINEERING MECHANICS FOR STRUCTURES, Louis L. Bucciarelli. This text explores the mechanics of solids and statics as well as the strength of materials and elasticity theory. Its many design exercises encourage creative initiative and systems thinking. 2009 edition. 320pp. 6 1/8 x 9 1/4. 0-486-46855-0

FEEDBACK CONTROL THEORY, John C. Doyle, Bruce A. Francis and Allen R. Tannenbaum. This excellent introduction to feedback control system design offers a theoretical approach that captures the essential issues and can be applied to a wide range of practical problems. 1992 edition. 224pp. 6 1/2 x 9 1/4. 0-486-46933-6

THE FORCES OF MATTER, Michael Faraday. These lectures by a famous inventor offer an easy-to-understand introduction to the interactions of the universe's physical forces. Six essays explore gravitation, cohesion, chemical affinity, heat, magnetism, and electricity. 1993 edition. 96pp. 5 3/8 x 8 1/2. 0-486-47482-8

DYNAMICS, Lawrence E. Goodman and William H. Warner. Beginning engineering text introduces calculus of vectors, particle motion, dynamics of particle systems and plane rigid bodies, technical applications in plane motions, and more. Exercises and answers in every chapter. 619pp. 5 3/8 x 8 1/2. 0-486-42006-X

ADAPTIVE FILTERING PREDICTION AND CONTROL, Graham C. Goodwin and Kwai Sang Sin. This unified survey focuses on linear discrete-time systems and explores natural extensions to nonlinear systems. It emphasizes discrete-time systems, summarizing theoretical and practical aspects of a large class of adaptive algorithms. 1984 edition. 560pp. 6 1/2 x 9 1/4. 0-486-46932-8

INDUCTANCE CALCULATIONS, Frederick W. Grover. This authoritative reference enables the design of virtually every type of inductor. It features a single simple formula for each type of inductor, together with tables containing essential numerical factors. 1946 edition. 304pp. 5 3/8 x 8 1/2. 0-486-47440-2

THERMODYNAMICS: Foundations and Applications, Elias P. Gyftopoulos and Gian Paolo Beretta. Designed by two MIT professors, this authoritative text discusses basic concepts and applications in detail, emphasizing generality, definitions, and logical consistency. More than 300 solved problems cover realistic energy systems and processes. 800pp. 6 1/8 x 9 1/4. 0-486-43932-1

THE FINITE ELEMENT METHOD: Linear Static and Dynamic Finite Element Analysis, Thomas J. R. Hughes. Text for students without in-depth mathematical training, this text includes a comprehensive presentation and analysis of algorithms of time-dependent phenomena plus beam, plate, and shell theories. Solution guide available upon request. 672pp. 6 1/2 x 9 1/4. 0-486-41181-8

Browse over 9,000 books at www.doverpublications.com